Forschungsberichte aus dem Lehrstuhl für Regelungssysteme

Technische Universität Kaiserslautern

Band 17

Forschungsberichte aus dem Lehrstuhl für Regelungssysteme

Technische Universität Kaiserslautern

Band 17

Herausgeber:

Prof. Dr. Steven Liu

Hengyi Wang

Delta-connected Cascaded H-bridge Multilevel Converter as Shunt Active Power Filter

Logos Verlag Berlin

Forschungsberichte aus dem Lehrstuhl für Regelungssysteme
Technische Universität Kaiserslautern

Herausgegeben von
Univ.-Prof. Dr.-Ing. Steven Liu
Lehrstuhl für Regelungssysteme
Technische Universität Kaiserslautern
Erwin-Schrödinger-Str. 12/332
D-67663 Kaiserslautern
E-Mail: sliu@eit.uni-kl.de

Bibliographic information published by the Deutsche Nationalbibliothek

The Deutsche Nationalbibliothek lists this publication in the Deutsche Nationalbibliografie; detailed bibliographic data are available on the Internet at http://dnb.d-nb.de .

ISBN 978-3-8325-5015-8
ISSN 2190-7897

Logos Verlag Berlin GmbH
Comeniushof, Gubener Str. 47,
10243 Berlin
Tel.: +49 (0)30 / 42 85 10 90
Fax: +49 (0)30 / 42 85 10 92
http://www.logos-verlag.de

Delta-connected Cascaded H-bridge Multilevel Converter as Shunt Active Power Filter

Multilevel-Umrichter aus kaskadierten H-Brücken in Dreieckschaltung als parallelkompensiertes aktives Leistungsfilter

vom Fachbereich Elektrotechnik und Informationstechnik
der Technischen Universität Kaiserslautern zur Verleihung
des akademischen Grades Doktorin der Ingenieurwissenschaft (Dr.-Ing.)
genehmigte Dissertation
von M. Sc. Hengyi Wang
geboren in JiangXi, V.R. China
D386

Tag der mundlichen Prüfung: 29.08.2019

Dekan des Fachbereichs: Prof. Dr.-Ing. Ralph Urbansky

Vorsitzender der Prüfungskommission: Prof. Dr.-Ing. Stefan Götz

1. Berichterstatter: Prof. Dr.-Ing. Steven Liu

2. Berichterstatter: apl. Prof. Dr.-Ing. Daniel Görges

Acknowledgement

This thesis presents the results of my work at the Institute of Control Systems (LRS), Department of Electrical and Computer Engineering, University of Kaiserslautern.

First and foremost, I would like to express my thanks and gratitude to Prof. Dr.-Ing. Steven Liu, the head of the Institute of Control Systems (LRS), for accepting me as a PhD candidate, the excellent supervision of my research, the scientific discussions and also the good researching atmosphere.

Further, I would like to thank apl. Prof. Dr.-Ing. Daniel Görges for the interest in my research and for joining the thesis committee as a reviewer. Thanks also go to Prof. Dr.-Ing. Stefan Götz for joining the thesis committee as a chair.

My time at the Institute of Control Systems is very enjoyable and rewarding. All the (former) colleagues created an open, cooperating and warm working atmosphere, such that I can enjoy my work so much. My thanks go to all them, especially to M. Sc. Alen Turnwald, M. Sc. Andreas Weißmann, M. Sc. Benjamin Watkins, M. Sc. Christoph Mark, apl. Prof. Dr.-Ing. Daniel Görges, Dr.-Ing. Fabian Kennel, Dr.-Ing. Felix Berkel, M. Sc. Filipe Figueiredo, M. Sc. Giuliano Costantini, M. Sc. Guihai Luo, Dr.-Ing. Guoqiang Li, M. Sc. Jawad Ismail, Dr.-Ing. Jianfei Wang, M. Sc. Kashif Iqbal, Dr.-Ing. Markus Bell, M. Sc. Markus Lepper, M. Sc. Min Wu, M.Sc. Muhammad Ikhsan, M. Sc. Pedro Dos Santos, Dipl.-Ing. Peter Müller, M. Sc. Qingshan Pan, M. Sc. Ramin Rostami, M. Sc. Sai Krishna Chada, Dr.-Ing. Sanad Al-Areqi, M. Eng. Sebastian Caba, Dr.-Ing. Stefan Simon, Dr.-Ing. Sven Reimann, Dr.-Ing. Tim Nagel, M. Sc. Tim Steiner, Dipl.-Ing. Tobias Lepold, M. Sc. Tobias Peschke, Dr.-Ing. Wei Wu, M. Sc. Xiang Chen, M. Sc. Xiaohai Lin, M. Sc. Yanhao He, M. Sc. Yanzhao Jia, Dr.-Ing. Yun Wan, M. Sc. Yakun Zhou, M. Sc. Zhuoqi Zeng and Priv. Doz. Dr.-Ing. habil. Christian Tuttas. Thanks also go to technicians Swen Becker and Thomas Janz and to the secretary Jutta Lenhardt for providing a good technical and administrative environment.

Finally, warm thanks from the deep of my heart to my parents and family for all the love and support over all the years. This thesis is dedicated to them.

Kaiserslautern, May 2019

Hengyi Wang

Abstract

This thesis addresses an optimal current operation strategy and harmonic interaction analysis of delta-connected cascaded H-bridge (CHB) multilevel converter based shunt active power filter (APF).

An optimal current operation strategy for a delta-connected CHB based shunt APF under non-ideal grid conditions is presented that minimizes the shunt APF apparent power and satisfies requirements on average power balance, power factor constraint, source current distortion and imbalance characteristics in compliance with grid codes. The optimization problem is formulated using a convex quadratic objective function and non-convex quadratic constraints and solved by sequential convex programming (SCP).

The presented harmonic interaction analysis is based on the derivation of a time-invariant model in *dq*-frame from a time-varying model in *abc*-frame. In the derived time-invariant model, the symmetrical components of each frequency order in the shunt APF variables can be decoupled while preserving the interaction with other variables. The harmonic interaction analysis technique is able to predict how harmonics propagate through the system and quantify the shunt APF variables.

The proposed current strategy and harmonic interaction analysis are evaluated by simulations in comparison with related approaches from literature, to demonstrate the effectiveness.

Contents

List of Figures XI

List of Tables XV

List of Symbols XVII

Acronyms XIX

1. Introduction 1

1.1. Motivation . . . 1
1.2. Objectives . . . 4
1.3. Related Work . . . 5
1.3.1. Current Operation Strategies . . . 6
1.3.2. Harmonic Interaction Analysis . . . 8
1.4. Outline and Contribution . . . 10
1.5. Publications . . . 12

2. Foundation of Delta-connected CHB Multilevel Converter 13

2.1. Voltage Source Converters . . . 13
2.1.1. Classification . . . 13
2.1.2. Applications . . . 18
2.1.3. Choice of Delta-connected CHB as Shunt APF . . . 23
2.2. Non-ideal Power Supplies . . . 24
2.2.1. Voltage Disturbances in Electric Power Systems . . . 24
2.2.2. General Description of Periodic Voltages in *abc* Frame . . . 26
2.3. Challenges of Operational Issues of Delta-connected CHB as Shunt APF under Non-ideal Power Supplies . . . 28

3. Optimal Current Operation Strategy 31
3.1. Definitions for Electric Quantity Measurement 31
3.2. Recommended Practice for Electric Power Systems 33
3.3. Operation Principle of Delta-connected CHB 35
3.4. The Proposed Strategy for Desired Terminal Currents and Circulating Current Calculation 39
3.4.1. Harmonic-based Distortion Analysis in dq Frame 40
3.4.2. Optimization Problem Formulation 44
3.4.3. Combination of Objective Function and Constraints 49
3.4.4. Solution of the Optimization Problem via Sequential Convex Programming 50
3.5. Sequential Convex Programming and Its Convergence 52
3.6. Block Diagram for Strategy Validation 54
3.6.1. Harmonic Detection Method 55
3.6.2. Modelling and Controller Design 56
3.6.3. Finite Horizon Discrete Time LQR via Dynamic Programming 59
3.7. Validation via Time-Domain Simulation 61
3.7.1. Case A 66
3.7.2. Case B 68
3.7.3. Case C 70
3.7.4. Case D 71
3.8. Discussion 73

4. Harmonic Interaction Analysis 83
4.1. Necessity of Harmonic Interaction Analysis 83
4.2. Time-varying Model and Properties 86
4.3. Observations of Power Flow and Branch Voltage Model 90
4.3.1. Foundation 90
4.3.2. Observations of Power Flow Model 91
4.3.3. Observations of Branch Voltage Model 95
4.4. Derivation of Time-invariant Representation for Power Flow 95
4.4.1. Time-invariant Representation for dc Component of Power Flow 96

4.4.2. Time-invariant Representation for Oscillating Component of Power Flow 96

4.5. Utilisation of Harmonic Interaction Analysis 98

4.5.1. Time-invariant Model Based on Harmonic Interaction Analysis 98

4.5.2. Determination of Control Variables 99

4.6. Validation via Time-Domain Simulation 103

4.6.1. Case A 106

4.6.2. Case B 108

4.6.3. Case C 108

4.6.4. Case D 110

4.6.5. Case E 110

4.7. Discussion 111

5. Summary and Outlook **123**

6. Zusammenfassung **127**

A. Supplementary Materials **131**

A.1. Power Component Calculation (Chapter 3) 131

A.2. Some Details of Eq. (4.26) 134

A.3. Harmonic Sequence Coupling 139

Bibliography **141**

List of Figures

2.1. Voltage source converter classification 14
2.2. A traditional two-level converter with IGBT as switches 15
2.3. (a)A three-level NPC (b)A three-level FCC 15
2.4. (a)half-bridge (b)H-bridge 16
2.5. (a)Single-star connected converter (b)Single-delta connected converter (c)Double-star connected converter (d)Triple-star connected converter (e)Double-delta connected converter (hexverter) 17
2.6. A general structure for distributed power systems with different distributed energy resources 20
2.7. An example of a modern grid 22
2.8. Basic circuits of standardized FACTS devices 22
2.9. Various arts of voltage disturbance 25
2.10. The positive-, negative- and zero-sequence component of an unbalanced three-phase voltage system of the n-th frequency order 28

3.1. A delta-connected CHB as shunt APF connected to the line 36
3.2. Block diagram of the optimal operation algorithm 39
3.3. Block diagram for the operation strategy validation: shunt connection of an active power filter to the line which feeds a nonlinear load. The active power filter applied here is delta-connected CHB, which is for simplicity represented by a H-bridge cell with a capacitor. . . 54
3.4. Feedback control loop for the delta-connected CHB system 59
3.5. Harmonic group according to IEC standard 61000-4-7 74

3.6. Case A with PHC (0.04 s-0.08 s), UPF (0.08 s-0.12 s), CST1 (0.12 s-0.16 s), and OPT (0.16 s-0.2 s). The APF injection at $t = 0.04$ s. From top to bottom are PCC voltages, source currents, branch currents, circulating current, the sum of SM capacitor voltages of each branch. 77

3.7. Case B with PHC (0.04 s-0.08 s), UPF (0.08 s-0.12 s), CST1 (0.12 s-0.16 s), CST2 (0.16 s-0.2 s), CST3 (0.2 s-0.24 s), and OPT (0.24 s-0.28 s). The APF injection at $t = 0.04$ s. From top to bottom are PCC voltages, source currents, branch currents, circulating current, the sum of SM capacitor voltages of each branch. 78

3.8. Case C with PHC (0.04 s-0.08 s), UPF (0.08 s-0.12 s), CST1 (0.12 s-0.16 s) and OPT (0.16 s-0.2 s). The APF injection at $t = 0.04$ s. From top to bottom are PCC voltages, source currents, branch currents, circulating current, the sum of SM capacitor voltages of each branch. 79

3.9. Case D with with PHC (0.04 s-0.08 s), UPF (0.08 s-0.12 s), CST1 (0.12 s-0.16 s), CST2 (0.16 s-0.2 s), CST3 (0.2 s-0.24 s), CST4 (0.24 s-0.28 s), and OPT (0.28 s-0.32 s). The APF injection at $t = 0.04$ s. From top to bottom are PCC voltages, source currents, branch currents, circulating current, the sum of SM capacitor voltages of each branch. . 80

3.10. Single-sided RMS amplitude spectrum of source currents, where I_{sjn} represents the RMS value of the nth order phase-j source current. . 81

4.1. Interaction of various quantities in one APF branch 84

4.2. Comparison of (a) linear system and (b) nonlinear system 85

4.3. Interaction example 94

4.4. Control diagram of the shunt APF system for harmonic interaction analysis validation 103

4.5. Case A with the PHC strategy (a) harmonic sequence amplitudes of branch powers in pu (base value 25 MVA) and (b) harmonic sequence amplitudes of switching functions in pu. 114

4.6. Case A with the PHC strategy: the sum of SM capacitor voltages in each branch (left) and switching functions (right) in time-domain 114

4.7. Case B with the PHC strategy (a) harmonic sequence amplitudes of branch powers in pu (base value 25 MVA) and (b) harmonic sequence amplitudes of switching functions in pu. 115
4.8. Case B with the PHC strategy: the sum of SM capacitor voltages in each branch (left) and switching functions (right) in time-domain 115
4.9. Case C with the PHC strategy (a) harmonic sequence amplitudes of branch powers in pu (base value 25 MVA) and (b) harmonic sequence amplitudes of switching functions in pu. 116
4.10. Case C with the PHC strategy: the sum of SM capacitor voltages in each branch (left) and switching functions (right) in time-domain 116
4.11. Case D with the PHC strategy (a) harmonic sequence amplitudes of branch powers in pu (base value 25 MVA) and (b) harmonic sequence amplitudes of switching functions in pu. 117
4.12. Case D with the PHC strategy: the sum of SM capacitor voltages in each branch (left) and switching functions (right) in time-domain 117
4.13. Case E with the PHC strategy (a) harmonic sequence amplitudes of branch powers in pu (base value 25 MVA) and (b) harmonic sequence amplitudes of switching functions in pu. 118
4.14. Case E with the PHC strategy: the sum of SM capacitor voltages in each branch (left) and switching function (right) in time-domain 118

List of Tables

3.1. Parameters used in optimization 62
3.2. Parameters used in simulations 62
3.3. Result summary of Case A 67
3.4. Result summary of Case B 68
3.5. Result summary of Case C 70
3.6. Result summary of Case D 72
3.7. APF power components of various strategies with unit of **MW** and **MVar** for active and reactive Power (Case A-B) 75
3.8. APF power components of various strategies with unit of **MW** and **MVar** for active and reacitve power (Case C-D) 76

4.1. Parameters used in simulations 104
4.2. Harmonic sequence amplitudes of capacitor sum voltages of Case A 119
4.3. Harmonic sequence amplitudes of capacitor sum voltages of Case B 119
4.4. Harmonic sequence amplitudes of capacitor sum voltages of Case C 120
4.5. Harmonic sequence amplitudes of capacitor sum voltages of Case D 121
4.6. Harmonic sequence amplitudes of capacitor sum voltages of Case E 122

A.1. Harmonic sequences in powers brought by the m-th branch voltage $\boldsymbol{u}^{abcm}$ and the k-th branch current $\boldsymbol{i}^{abck}$, where $m > k$ 139
A.2. Harmonic sequences in powers brought by the m-th branch voltage $\boldsymbol{u}^{abcm}$ and the k-th branch current $\boldsymbol{i}^{abck}$, where $m < k$ 139
A.3. Harmonic sequences in powers brought by the m-th branch voltage $\boldsymbol{u}^{abcm}$ and the k-th branch current $\boldsymbol{i}^{abck}$, where $m = k$ 139

List of Symbols

Matrices

$\mathbf{0}$	Zero matrix
$\boldsymbol{I}$	Identity matrix

Parameters

ω	Fundamental angular frequency
C	APF submodule capacitance
C_{sum}	Equivalent APF branch capacitance defined as $C_{sum} = C/N$
N	Number of submodules per APF branch
r, L	APF branch resistance and inductance

Superscripts

$*$	Indicates reference values
$\top$	Indicates transpose
abc	Natural three-phase coordinates
$abc+n$	The nth positive-sequence component in natural three-phase coordinates
$abc-n$	The nth negative-sequence component in natural three-phase coordinates
$abc0n$	The nth zero-sequence component in natural three-phase coordinates

$abcn$	The nth frequency component in natural three-phase coordinates
dq	Synchronous coordinates
$dq+n$	Synchronous frame rotating at $n\omega$ for the positive-sequence component
$dq-n$	Synchronous frame rotating at $n\omega$ for the negative-sequence component
$dq0n$	Synchronous frame rotating at $n\omega$ for the zero-sequence component

Variables

i_0	APF circulating current
i_{ab}, i_{bc}, i_{ca}	APF branch current at branch-ab, bc, ca
i_{La}, i_{Lb}, i_{Lc}	Load current at phase-a, b, c
i_{sa}, i_{sb}, i_{sc}	Source current at phase-a, b, c
p_{ab}, p_{bc}, p_{ca}	APF branch power at branch-ab, bc, ca
s_{ab}, s_{bc}, s_{ca}	APF modulation signal at branch-ab, bc, ca
u_{ab}, u_{bc}, u_{ca}	APF branch voltage at branch-ab, bc, ca
u_{Cab}, u_{Cbc}, u_{Cca}	APF submodule capacitor sum voltages in branch-ab, bc, ca
v_{sab}, v_{sbc}, v_{sca}	PCC line-line voltage at phase-ab, bc, ca
v_{sa}, v_{sb}, v_{sc}	PCC voltage at phase-a, b, c
w_{ab}, w_{bc}, w_{ca}	APF submodule capacitor sum energies in branch-ab, bc, ca

Acronyms

APF	Active Power Filter
BESS	Battery Energy Storage System
CHB	Cascaded H-bridge
DER	Distributed Energy Resource
DESS	Distributed Energy Storage System
DVR	Dynamic Voltage Restorer
FACTS	Flexible AC Transmission System
FCC	Flying Capacitor Converter
HVDC	High Voltage Direct Current
KKT	Karush-Kuhn-Tucker
LQR	Linear Quadratic Regulator
LSPWM	Level-Shifted PWM
M3C	Modular Multilevel Matrix Converter
MMC	Modular Multilevel Converter / Modularer Multilevel Umrichter
NPC	Neutral Point Clamped Converter
PCC	Point of Common Coupling

PHC	Perfect Harmonic Cancellation
PSPWM	Phase-Shifted PWM
PV	PhotoVoltaic
QCQP	Quadratic Constraints Quadratic Programming
RLT	Reformulation-Linearization Technique
RMS	Root Mean Square
SCP	Sequential Convex Programming
SDP	Semidefinite Programming
SM	SubModule
STATCOM	Static Synchronous Compensator
TDD	Total Demand Distortion
THD	Total Harmonic Distortion
UPF	Unity Power Factor
UPFC	Unified Power Flow Controller
VSC	Voltage Source Converter
VUF	Voltage Unbalance Factor

1. Introduction

1.1. Motivation

Nowadays the wide utilisation of power-electronics-based loads causes distortion in voltages and currents as well as reactive power components. These electrical power quality problems have become an important issue in transmission and distribution power systems. Active Power Filters (APFs) are filters that can perform the job of harmonic and reactive power elimination. Depending on the power quality problem to be solved, APFs can be implemented as shunt type, series type, or a combination of shunt and series types. Shunt APF is the use of an electronic power converter to compensate reactive and harmonic currents so that the source currents after compensation can meet grid codes. Shunt APF development/design including several considerations such as follows:

- converter topology
- converter operating current strategy
- modelling and harmonic interaction analysis

A very common shunt APF configuration in three-phase systems is traditional Voltage Source Converters (VSCs) shown in Fig. 2.2. However, limited by voltage ratings of semiconductor switches, this topology is not suitable for medium- or high-power applications. As a result, multilevel converters, including delta- and star-connected Cascaded H-bridge (CHB) multilevel converters [WSDC15, YSL+17], Neutral Point Clamped Converters (NPCs) [MMT+17], Flying Capacitor Converters (FCCs) [AJKM16], Modular Multilevel Converters / Modulare Multilevel Umrichter (MMCs) [SLZ+16] and among others, which can be directly connected to medium voltage systems without transformers are emerged to accommodate this situation. Each converter topology represents advantages and disadvantages. The CHB multilevel converter has the advantage to reduce harmonic distortion level and the rating of power switches. The delta-connected CHB makes the negative-sequence reactive power compensation possible due to a circulating current flowing inside. However, it needs a high number of dc-link capacitors and there is an imbalance control problem, increasing the control complexity. Due to the strong and weak points associated with each topology, the selection depends on particular design criteria. This dissertation is focused on shunt APF based on delta-connected cascaded H-bridge multilevel converter, which has found widespread use in practical projects.

The choice of the converter topology is part of the full shunt APF system. Another issue concerned to shunt APFs is the reference operating current calculation. The reference operating current calculation for delta-connected CHB-based shunt APF includes two parts: a) compensation/terminal current calculation; b) circulating current calculation. The present methods calculate terminal currents and circulating current separately instead of as an integrity. All terminal current calculation methods require that the average power delivered by the source equals to that consumed by the load, in other words, the average value of the total APF active power is zero. The traditional Perfect Harmonic Cancellation (PHC) and Unity Power Factor (UPF) strategies are well-known terminal current calculation methods. The PHC strategy aims at harmonics and unbalance free source currents. The UPF strategy results in source currents in phase with Point of Common Coupling (PCC) voltages. PHC and UPF lead to the same result when the PCC voltages contain solely the positive-sequence fundamental frequency component. However, when the PCC voltages contain harmonics and/or unbalance, the PHC power factor may be low while the UPF source currents may violate grid codes. Optimal algorithms have been proposed in some works to compromise PHC and UPF to maximize power factor while meeting grid codes. One common feature of all the reference terminal current calculation methods is that they do not take the internal structure and characteristics of each converter topology into account, leading to the same results for each converter. Looking into the delta-connected CHB multilevel converter, one key phenomenon that is specific to this converter and provides opportunities for unique design and control optimization is the circulating current, which flows in the delta configuration in such a way that it does not influence the total APF active power but is able to redistribute the active power among the three branches to maintain SubModule (SM) capacitor voltages, when the average active power released and absorbed by per-branch is balanced. Circulating current calculation for objective minimization/maximization such as SM capacitor voltage ripple minimization has been researched, with the foundation that the terminal currents are determined by the conventional PHC or UPF strategy. In fact, the joint calculation of terminal currents and circulating current can be considered to further minimize/maximize objectives, but has not yet been discussed in present literatures. The operational issue concerned to the delta-connected CHB-based shunt APF, a joint reference current calculation, out of the consideration of grid code requirement and APF apparent power minimization, will be tackled in this dissertation.

The investment and operational cost of a shunt APF system should also be taken into account. Conventionally shunt APFs employ very large electrolytic capacitors so that the capacitor voltage ripple is negligibly small (typically less than $\pm5\%$). However, such capacitors are known to be bulky, weighty and costly. The SM dc-link application with small electrolytic capacitors or even film capacitors that have lower capacitance than electrolytic type capacitors with the same volume, leads to higher amplititude low-frequency ($<2\,\mathrm{kHz}$) capacitor voltage ripple, however, can achieve lower cost and size [FHA15, WB14]. In addition, for the same switch and dc-link capacitor in each SM, the higher number of SMs is accompanied with the higher quality of the output voltage, however, the total number of devices, conduction loss which is a function of the SM number inserted in the conduction path, and converter cost are also higher. Therefore, a minimum number of SMs and each SM capacitance with a reduced value, are desired to reduce the converter volume, losses, cost of cascaded multilevel converters [SA15, LQTH16]. The shunt APF modelling and analysis can be split into two subsystems: a) the ac-side dynamics; b) the dc-side dynamics. As for the subsystem a), they are traditionally modelled as linear equations by approximation of SM capacitor voltages as constant, which is reasonable when APFs employ very large electrolytic capacitors. It is useful to use Park's transformation using reference frames rotating at fundamental and harmonic frequencies. This makes it possible to see the d and q components of the converter branch currents/voltages as dc component at each frequency order. However, such linear equations for the ac-side dynamics are inaccurate when SM capacitor voltage harmonics are large, which happens when small electrolytic capacitors or even film capacitors are applied, resulting in that the ac-side dynamics cannot be approximated as a linear system. As for the subsystem b), the harmonics in the instantaneous APF branch power which is the cross coupling of the APF branch voltage and current are finally reflected in the SM dc-link capacitors. There are very few studies discussing the relationship between the harmonics in the APF branch powers and SM capacitor voltages. This dissertation has proposed a harmonic interaction analysis tecnnique that is based a state-space model in dq frame, which is derived from a mathematical abc frame model, in which the ac- and dc-side APF dynamics are integrated. The description of the ac-side dynamics does not ignore the coupling of capacitor voltage ripple and switching functions. The expression of the dc-side dynamics considers the coupling of APF branch voltages and currents. The accurate relationship between the harmonics in the APF branch powers and SM capacitor voltages is given. The proposed harmonic interaction analysis, which is based on the derived dq frame model, is able to predict how harmonics propagate through the system and quantify electrical and nonelectrical quatities (switching functions), providing great reference for APF understanding, designing and controlling.

1.2. Objectives

In this dissertation two main objectives in the operation of the delta-connected CHB-based shunt APF can be summarized as:

Current operation strategy An optimal strategy is presented to determine the combined control reference for the terminal currents and the circulating current of the delta-connected CHB-based shunt APF, hereby minimizing the operational power under ideal and non-ideal PCC voltages conditions, where the ideal PCC voltages contain only the positive-sequence fundamental frequency component and the non-ideal PCC voltages contain unbalance and/or harmonic components. The optimization is carried out to simultaneously meet the desired source current distortion limits, the source current imbalance characteristics and power factor as well as power balance requirements. This strategy cannot be directly used when the power systems contain interharmonics, such as Fig. 2.9. (a) and (c). Reasons will be given in Section 3.8.

Harmonic interaction analysis The proposed harmonic interaction analysis is based on the derivation of a time-invariant representation of an integrated model of ac- and dc-side dynamics of the delta-connected CHB - based shunt APF, achieving accurate variable quatification. Retaining the cross coupling between various variables, the time-invariant representation is a multi-input multi-output (MIMO) system, which is characterized by a nonlinear relationship between inputs and state-variables/outputs. Take the ac-side dynamics as an example, the nth frequency order in switching functions can arouse converter currents/voltages with not only the nth but also other frequency orders, while conventionally the relationship between the switching functions and converter currents/voltages is linear by assuming SM capacitor voltages as constant. The analysis technique is proposed when the power systems contain the fundamental frequency and/or harmonic components possibly with unbalance, and even applicable if a couple of interharmonics are of concern.

It should be noticed that although the two objectives implemented in Chapters 3-4 are validated in steady-state (or continuous) power systems, they can be still applied when some types of disturbances (temporarily or permanently) happen, such as Fig. 2.9. (e), as long as the disturbance duration is longer than one fundamental period and the frequencies contained in the power systems are integral multiple times of the fundamental frequency.

1.3. Related Work

After the objectives are specified, the related work is given. Thereby, the open points, which are covered in the thesis, are identified. First, the current operation strategies for three-phase shunt APFs are reviewed. Second, the harmonic interaction analysis methods are discussed.

1.3.1. Current Operation Strategies

The joint design of terminal currents and circulating current simultaneously for delta-connected CHB-based shunt APF for power consumption reduction, to author's best knowledge, has not been studied so far. However, a review of previous work is still given, which focuses on how to calculate either the terminal currents or the circulating current.

Strategies to determine the terminal currents

The current strategies which calculate the terminal currents, have nothing to do with the inner structure of the shunt APF itself, only there is a difference between three-wire and four-wire systems, since three-wire systems cannot compensate the zero-sequence current from the loads while four-wire systems can [VEToMm06]. When the inner converter structure is not considered, possibly additional freedom degrees in some structures are not taken into account that can be actually used to further improve the system performance like the circulating current in delta-connected CHB. The terminal current strategies can be classified into two categories, namely *the traditional current strategies*, such as the generalized p-q, i_d-i_q, the PHC and UPF strategy, and *a joint consideration of traditional strategies*, for instance the joint consideration of the PHC and UPF strategy.

Traditional Current Strategies Some current strategies have been presented in [MCG07, RMG08] for the desirable terminal currents calculation: the generalized p - q strategy, the i_d - i_q method, PHC and UPF strategy respectively. In all of the current strategies, the constant active power provided by the source is equal to that consumed by the load. Under the p - q strategy, the load current compensation is achieved by cancellation of the instantaneous zero-sequence power at the source side. The i_d - i_q method is based on the assumption that the source provides the mean value of the direct-axis component of the load current. In the PHC strategy, the source currents are in phase with the fundamental positive-sequence PCC voltages. In this strategy, the imbalance and all the harmonics are compensated. Under the UPF strategy, the waveform of the source currents resembles that of the PCC voltages, which means, the three-phase load current is compensated for the purpose that the nonlinear load together with the compensator are viewed as a symmetrical and constant resistive load from the point of view of the PCC voltages. If the PCC voltages are ideal, containing the fundamental positive-sequence phase component, PHC and UPF can be achieved simultaneously. These two goals cannot be carried on when the PCC voltages are distorted and/or unbalanced. For instance, if the PCC voltages are severely distorted, the Total Harmonic Distortion (THD) of the UPF source currents may exceed the limitation of the standard requirement although the power factor is unity (in three-wire systems). The PHC source currents contain neither harmonics nor unbalance, however, the PHC efficiency can be low under distorted and/or unbalanced PCC voltages.

A joint consideration of traditional strategies: Under non-ideal grid conditions, a joint design of PHC and UPF to maximize the power factor by meeting some current harmonic distortion constraints has been reported based on a rather simple combination [RTGG01, UMG09, KKZ13]. References [RTGG01, UMG09] formulate the optimization problem as nonlinear and solve the formulated problem with the MATLAB optimization toolbox. In reference [KKZ13] a single-step noniterative optimized control algorithm has been proposed for a three-phase four-wire shunt APF to achieve a compromization between efficiency and total harmonic distortion. In this paper [KKZ13] the comparison of the noniterative optimized control algorithm and the iterative Newton-Raphson (NR) method shows that the proposed single-step noniterative algorithm has smaller computation time. The convergence analysis has not been done in literatures [RTGG01, UMG09, KKZ13].

Strategies to determine the circulating current

The strategies which calculate the circulating current, which flows within converters and does not affect the terminal currents injected at the grids, have been proposed depending on different objectives, for instance circulating current suppression, selective harmonic current suppression [WHO14, YWT+18], capacitor voltage ripple suppression [WLZX13, KWS10], converter current rating reduction [BMLB15a, LZX+16, KKGB15], power semiconductor temperature fluctuation minimization [BMLB15b], among others. References [WHO14, YWT+18] claim that the MMC circulating current can increase semiconductor current stress, converter power losses, submodule capacitor voltage ripples and even instability during transient state, they made effort to perfectly suppress even-order harmonics in the differential/circulating current. However, for some special applications such as capacitor voltage ripple shaping, converter branch current Root Mean Square (RMS) reduction, and so on, the circulating current harmonics are not necessarily undesirable. Circulating current injection methods have been proposed for capacitor voltage ripple suppression [WLZX13, KWS10]. These methods result in the usage of smaller capacitors and reduced costs. Current RMS reduction via circulating current injection has also been investigated for MMCs in High Voltage Direct Current (HVDC) applications/driver systems [BMLB15a, LZX+16, KKGB15] so that the device losses can be reduced. Also the effect of the circulating current on minimizing the temperature fluctuations of power semiconductors in an MMC has been studied in [BMLB15b], resulting in the improvement the devices life time.

The basic idea to inject the fundamental frequency circulating current for delta-connected CHB-based Static Synchronous Compensators (STATCOMs) under unbalanced and undistorted grid conditions has been reported in [BB17, HMA12, YKT+17, WCCC17] for power balancing among the converter branches. In these literatures, the circulating current is hardly adjusted since there is only fundamental-frequency component in the power supplies and load currents. [WSDC15, BJH+15] discuss circulating current injection for branch current RMS reduction of the delta-connected CHB. The approaches unfortunately can only work under ideal power supplies.

1.3.2. Harmonic Interaction Analysis

So far the harmonic interaction analysis of delta-connected CHB-based shunt APFs has not been reported, however, references [IANN12, JJ15, SP04] have proposed the analysis with frequency domain models for MMCs [IANN12, JJ15] and CHB-based STATCOMs [SP04], where the desired/nominal low-frequency switching function is assumed as the ratio between the desired SM output voltage and the dc component of the SM capacitor voltage, leading to that the frequency orders in the switching function are the same with those in the desired SM output voltages, thus the analysis results are relatively simple and incomplete. The similar assumption has been adopted in references [WCC17, CWL+15, LWY+17, KV17, DXM16], which have studied CHB multilevel converters. Actually, due to the existing harmonics in the SM capacitor voltages, the switching functions contain other low-frequency orders besides those in the desired SM output voltages.

References [SL14, MH09, KWB+17, KWB+16] have proposed harmonic interaction analysis based on that the modulation signals contain only the fundamental frequency which is calculated from the conventional method as in [SP04], and the switching function harmonics are brought during the modulation stage. Therefore different modulation strategies lead to different analysis results. In such methods, when the fundamental frequency modulation signal and the modulation strategy are settled, the mathematical expression of the switching function which includes baseband harmonics (i.e., simple harmonics of the fundamental line-frequency), harmonics of the carrier frequency and carrier sidebands which cluster around the carrier harmonics can be immediately obtained. Based on the expression of the switching function, they start the analysis step by step in the time-domain [SL14] or the frequency domain with the help of the Fourier transform and the convolution theorem [MH09, KWB+17, KWB+16].

There are two common features of [IANN12, JJ15, SP04, SL14, KWB+17, KWB+16]. One is that they have not considered the low-frequency harmonic orders in switching functions which are caused by the interaction with SM capacitor voltage harmonics. Another feature is that they deal with each phase separately, therefore there is no limitation for the system as it can be either single- or poly-phase. However, the decoupling of the sequence quantities, which is essential to analyse the three-phase systems with unbalance, has not been directly achieved.

References [BDSD18, JJ16] successfully achieve decoupling the sequence quantities of each frequency order when dealing with the harmonic coupling for MMCs in multiple *dq* rotating coordinate frames, including dc, the 1st to 3rd harmonics in the positive-, negative- and zero-sequence. However, it has not been extended to a high number of harmonics and a systematic approach has not been developed so that it cannot be applied to a network which contains higher frequency orders.

1.4. Outline and Contribution

This dissertation is consisted of two parts. The first part focuses on an operation strategy to calculate optimal terminal currents and circulating current (Chapter 3) and the second part (Chapter 4) is devoted to harmonic interaction analysis.

Chapter 2 The application of various voltage source converters is reviewed. The reasons to choose delta-connected CHB multilevel converter as shunt active power filter are given. In addition, the operation of delta-connected CHB as shunt APF operated under non-ideal PCC voltages, for a generic consideration, will be discussed in Chapters 3-4, thereby various faults in power supplies are reviewed and the mathematical expression of non-ideal PCC voltages is given.

Chapter 3 An optimal current operation strategy for a delta-connected CHB-based shunt APF under non-ideal grid conditions is presented that minimizes the APF apparent power and satisfies requirements on average power balance, power factor constraint, the source current distortion constraint as per IEEE STD-519 and imbalance characteristics as per IEEE STD-1159. This optimal strategy consists of two parts: the optimization problem formulation and the optimal solution searching for the formulated problem. The design approach is explained step by step including the appropriate analysis to formulate an optimization problem, which has a quadratic convex objective function and quadratic non-convex constraints, called non-convex Quadratic Constraints Quadratic Programming (QCQP) problem. In order to solve the non-convex QCQP problem, a proper treatment of the non-convex optimization based on local linearisation and iterative sequential programming have been made, resulting in the convergence to a local optimal solution. The operational power comparison of the presented optimal strategy, the traditional UPF and PHC strategy combined with the fundamental frequency circulating current, and the current strategies with feasible solutions, which are slightly shifted from the optimal solution of the non-convex QCQP problem validates the minimum shunt APF apparent power in the presented optimal strategy.

The presented strategy makes a contribution in the context of a joint reference current calculation, including the terminal currents and the circulating current in the delta-connected CHB-based shunt APF. Optimal strategies in the present research, typically propose the calculation of either the optimal terminal currents [RTGG01, UMG09, KKZ13] or the optimal circulating current [WSDC15, BJH+15]. Additionally, this chapter makes a contribution to the problem formulation. This chapter uses some concepts in [RTGG01, UMG09, KKZ13], where the concepts without transformation are directly formulated into optimization problems, resulting in nonlinear problems in a manner reflecting the situation being modelled but cannot be classified into one specific category of optimization programming. For such formulated problems without appropriate transformations, stochastic methods that cannot guarantee the convergence to an optimal solution in finite time are often used to solve these optimization problems. In the dissertation some proper transformation of the concepts is made targeting for a good problem formulation that the existence of a unique and optimal solution is assured.

Chapter 4 The harmonic interaction analysis is presented to predict the harmonics propagated through the system and quantify the electrical and non-electrical variables (switching functions/modulation signals) for the delta-connected CHB based shunt APF, based on the assumption that the switching functions are unknown beforehand. The main difficulty of this part is to derive the time-invariant dq representation from a time-varying expression, denoted by $x^{abc} = y^{abc} \circ z^{abc}$, in which $\boldsymbol{x}^{abc} = [x_{ab} \quad x_{bc} \quad x_{ca}]^\top$, $\boldsymbol{y}^{abc} = [y_{ab} \quad y_{bc} \quad y_{ca}]^\top$ and $\boldsymbol{z}^{abc} = [z_{ab} \quad z_{bc} \quad z_{ca}]^\top$ are three-phase time-varying signals containing multiple frequencies with the positive-, negative- and zero-sequence components. In addition, $\circ$ denotes element multiplication, i.e., $\begin{bmatrix} a \\ b \end{bmatrix} \circ \begin{bmatrix} c \\ d \end{bmatrix} = \begin{bmatrix} ac \\ bd \end{bmatrix}$. The electrical quantities (currents, powers, SM capacitor voltages) and switching functions can be quantified based on the harmonic interaction analysis and the derived time-invariant representation. Comparisons of variable quantification between with and without the harmonic interaction consideration have been made, validating the accuracy of the presented harmonic coupling analysis results.

The presented analysis makes a contribution in the context of the time-invariant derivation in which the symmetrical components at each frequency order can be decoupled while preserving the interaction with other variables. Therefore this work is specified for three-phase systems, unlike the harmonic interaction analysis which tackles each phase separately and have not decoupled sequence quantities [IANN12, JJ15, SP04, SL14, KWB+17, KWB+16]. In addition, the presented harmonic interaction and time-invariant model make a contribution of the steady-state quantification at the low-frequency range for the delta-connected CHB-based shunt APF.

1.5. Publications

A modelling and control method of the delta-connected CHB as STATCOM is introduced in [WTWL14]. [WL15] proposes an adaptive Kalman filter for harmonic detection. The separate calculation of an optimal reference terminal current and the circulating current can be seen in [WL16] and [WL17] respectively. A joint design of terminal currents and circulating current for APF apparent power minimization is introduced in [WL19b]. The harmonic interaction analysis has been published in [WL19a].

2. Foundation of Delta-connected CHB Multilevel Converter

The chapter reviews the classification and application of various voltage source converters, including multilevel converters. The reasons to choose delta-connected CHB as shunt APF are presented. Various power system disturbances are introduced, as the preliminary knowledge for the future chapters. In addition, the challenges of the two tasks of this dissertation are discussed.

2.1. Voltage Source Converters

2.1.1. Classification

The classification of voltage source converters (VSCs) is various dependent on different criterion, such as the number of phases, types of semiconductor devices, topologies and etc. Based on the number of phases to classify VSCs it can be such as two-wire (single-phase) and three- or four-wire three-phase systems. With the development of power systems, in order to meet the demand of high power, cost reduction and efficiency, at one side, the researchers have reacted to develop semiconductor technology to reach higher nominal voltages and currents, at the other side, new converter topologies have been developed, known as multilevel converters [RFK+09]. The advantages of multilevel converters lie in less harmonic generation, low switching losses, higher voltage operating capability compared with traditional two-level VSCs. Thus for high-power applications, such as transmission and distribution voltage levels, the multilevel-based VSCs become a more attractive solution. Some multilevel configurations have been studied and documented in technical literature, such as NPC, FCC and other modular multilevel VSCs based on cascaded H-bridge or cascaded half-bridge with isolated dc-link supply.

Fig.2.1 based on the structures shows the classification of VSCs for overview.

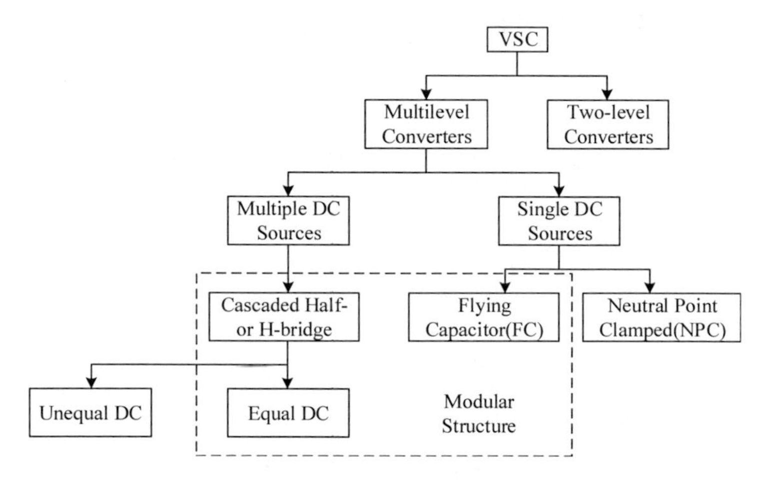

Figure 2.1.: Voltage source converter classification

Traditional Voltage Source Converters

Fig. 2.2 shows a traditional two-level converter. Because of the well-known circuit structure and control methods, two-level VSCs are competitive for low-voltage applications, up to few kV. In the case of the conventional two-level converter configurations, by going to higher power application, higher rated semiconductors are needed, which are more expensive. The harmonic reduction of this converter output is achieved by raising the switching frequency, leading to higher power losses. If other power quality requirements have to be fulfilled, then the need of power filters is introduced.

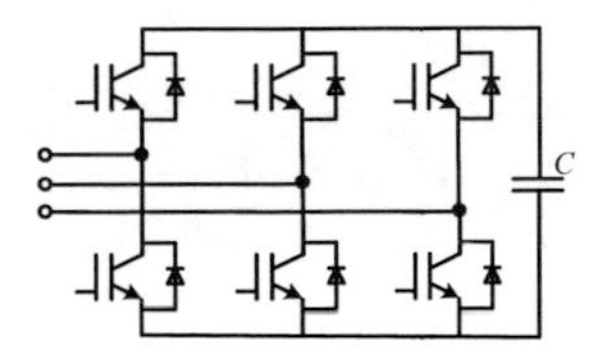

Figure 2.2.: A traditional two-level converter with IGBT as switches

Conventional Multilevel Converters

The following are those most known multilevel converter topologies:

1. Neutral Point Clamped Converter (NPC), a three-level NPC is shown as Fig. 2.3.(a)
2. Flying Capacitor Converter (FCC), a three-level FCC is shown as Fig. 2.3.(b)

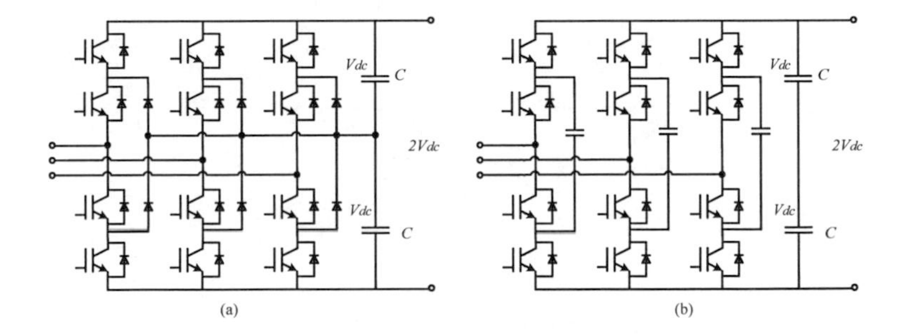

Figure 2.3.: (a)A three-level NPC (b)A three-level FCC

In 1980s, NPC which produces multilevel output waveforms was firstly introduced by Baker [Bak80]. After this achievement for over a decade, researchers put their effort on this topology and its controlling methods. One of the operation issues for NPC is dc-link capacitor voltage balancing [PSMD14]. NPCs are very competitive for medium-voltage applications (several kV) but do not scale well to many levels.

In 1990s, another topology, FCC, was developed, which has a modular structure and can be more easily extended to achieve more voltage levels and higher power ratings. The addition of power cells provides more redundant switching states, compared with NPCs. This redundancy is very important for safe operation and voltage balancing, but results in requiring more complex controlling methods.

Modular Multilevel Cascade Converters

Modular multilevel cascade converters are converter structures constituted by several branches, each of them being made of series-connected strictly identical submodules and reactance.

Submodule Structure Modular multilevel cascade converters employ strictly identical submodules to all its members. Fig. 2.4.(a) and (b) are two common submodule structures: half-bridge and H-bridge. The H-bridge submodule is built with four switches and a capacitor in parallel, this configuration makes a negative voltage output possible.

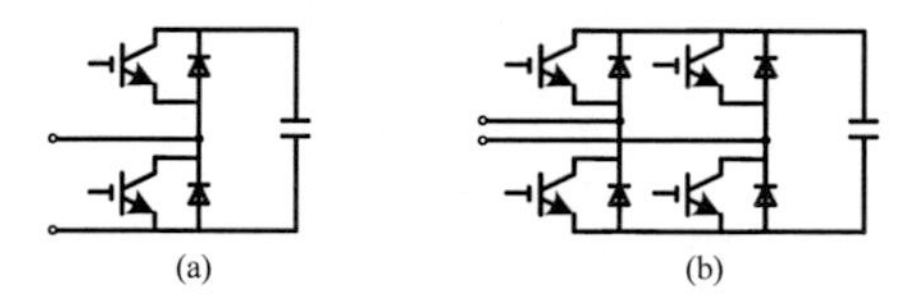

Figure 2.4.: (a)half-bridge (b)H-bridge

Topologies of Modular Multilevel Cascade Converters There are topologies such as single-star and single-delta connected converter, double-star connected converters, triple-star connected converter and hexverter for ac/ac conversion. Fig. 2.5 shows the topologies of them:

1. Single-star and single-delta connected converter, Fig. 2.5.(a) and (b) respectively. The structures are based on H-bridge as Fig. 2.4.(b). The main application is reactive power compensation and active power control.

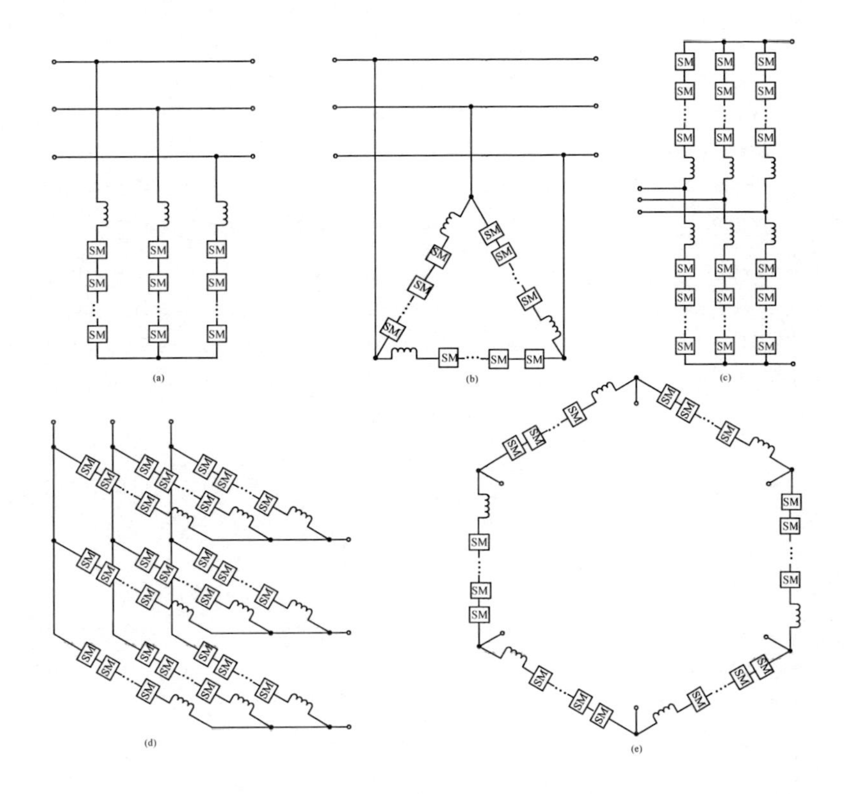

Figure 2.5.: (a)Single-star connected converter (b)Single-delta connected converter (c)Double-star connected converter (d)Triple-star connected converter (e)Double-delta connected converter (hexverter)

2. Double-star connected converter (usually called MMC) as shown in Fig. 2.5.(c). There are two situations dependent on the submodule structure: (1) rectification and inversion [LCM16] when half-bridge as Fig. 2.4.(a) is applied; (2) three- to single-phase frequency changer [WKS10] such as for railway supply applications in the case of H-bridge.

3. Triple-star connected converter (also as Modular Multilevel Matrix Converter (M3C)) [MFYI17, KHA14, EAN01] and double-delta connected converter (hexverter) [KBM15, BM13], shown in Fig. 2.5.(d) and (e). Both converters are based on H-bridge, such topologies are applied to connect two three-phase systems of different frequencies, voltage amplitudes and load angles.

The topology of single-delta connected converter is researched in this work. Different researchers use different names for above modular multilevel cascade converters [Kol14]. Here single-star and single-delta connected converter of Fig. 2.5.(a) and (b) are respectively named as star- and delta-connected CHB multilevel converter.

2.1.2. Applications

Voltage source converters have found an important market penetration in residential and industrial applications, motor drive applications and transportation systems such as ship propulsion and high-speed train traction, energy and power systems.

Residential and industrial applications: The application of VSCs can be found wide presence in the residential buildings such as space heating, air conditioning, lighting and so on. According to [MUR03], by application of converters for adjustable speed in the heat pump systems for space heating and air conditioning, the electric energy consumed can be reduced to 70% of that in conventional single speed heat pumps. An example of the application of traditional VSC of back-to-back structure is proposed in [TDMMN15] under the low voltage grid for heat pump systems.

Another interesting field of industrial application of VSCs covers such as cement industry, steel rolling mills, welding and induction heating [MUR03]. A back-to-back NPC have been proposed to meet the requirement of variable frequency in mining, cement, oil and gas industry [SMSG+10]. For the driven system of steel industry under medium voltages, the authors of [NFCP+17] have compared NPC, MMC and M3C. The literature [NSBP17] proposes a welding power supply system, where the dc/dc converter is based on H-bridge, to possess a constant dc voltage with current limiting feature.

Motor drive applications: The application of VSCs to driven systems is operated in a wide power range, from several watts to many thousands of kilowatts. The author in [Lor01] figures out the importance of power electronic converters in robotics. Here some examples of transportation drive are also given. Various converter structures in [KTHC16] have been reported to supply the motors for ac railway electric traction systems of medium voltages, for instance, 25 kV with 50 Hz and 15 kV with 16.7 Hz. The application of multilevel converters such as NPC and cascaded H-bridge for medium voltage shipboard power systems has also been introduced in [TRP+15].

Energy and power systems: Voltage source converters can be applied as regenerative converters, HVDC transmission devices, Distributed Energy Storage System (DESS) and Flexible AC Transmission System (FACTS) equipments for increased power quality and efficiency in energy and power systems.

The FCC, NPC, and MMC are presented in [FAD09] for HVDC applications. It points out that MMC seems to be more suitable than others for different number of voltage levels.

A grid-connected device for electricity storage such as PhotoVoltaic (PV) systems and wind energy conversion can be classified as a Distributed Energy Resource (DER) system, and is called a Distributed Energy Storage System (DESS) [Sin09, BTLT06]. By means of an interface, DER systems can be managed and coordinated within the utility grid. Fig. 2.6 shows a general structure for a distributed power system with different distributed energy resources, in which hydro and wind power plants output ac voltages, PV, batteries and fuel cells are with dc voltage output.

Hydropower takes up by far the largest area to all renewable electricity. Two-level VSCs and NPCs are compared in [VDB+18] for hydropower plants. The next largest share of renewable power was provided by wind power. 7.5 MW turbines are the largest today for wind energy conversion according to the european wind energy association. The NPC in back-to-back configration is applied for a small-scale (rated at 5 MW) in [MTMT10] and a large-scale (rated at 300 MW) in [XYS07] wind farms respectively to supply to 66 kV and 400 kV utility grid though a step-up transformer. The M3C is researched in [DCE+17] to drive multil-MW wind energy conversion systems. After hydro and wind power, PV is the third most important renewable energy in terms of globally installed capacity. Until now out of technical considerations it was adopted only on the high voltage levels for large wind power plants, yet recently more and more PV plants installed and integrated mainly to low and medium voltage grids [CKST12]. Traditional two-level VSCs and NPCs are applied in [KHS15] for grid integration of PV power, while delta-connected CHB is as the interface in [SAT17, YKT+17].

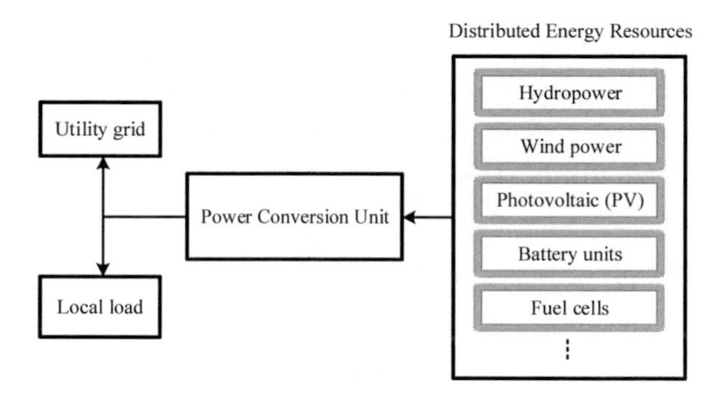

Figure 2.6.: A general structure for distributed power systems with different distributed energy resources

Battery Energy Storage Systems (BESSs) [LSN+14] are used for multiple applications, such as wind and solar power smoothing, peak shaving, frequency regulation, electric vehicle charging stations and other applications [QZLY11]. Further, the energy storage systems are needed to develop the microgrids and the future smart grid. Modern battery energy storage systems are based on the combination of a multilevel converter with an advanced battery technology, such as lithium (Li)-ion, sodium sulphur (NaS), nickel metal hydride (NiMH), and so on. When the grid demands power to be supplied from the batteries, the system chooses the optimal protocol for releasing charge while accounting for both the current state of the batteries and the grid's demand request. The MMC is used as the interface between a high-power battery energy storage system and a medium-voltage grid in [Bie13]. Fuel cells are also well used for distributed generation applications, and can essentially be described as batteries which never become discharged as long as hydrogen and oxygen are continuously provided. A tradition three-phase two-level VSC and a NPC is respectively controlled in [RKM18] and [HKE14] as the interface to a low and medium voltage grid.

Fig. 2.7 shows an example of a modern grid, from power generation, transmission and distribution systems to the end point of energy usage. The increasing penetration of renewable sources of energy, semiconductor based electronic equipment, nonlinear loads, data centres, industries running on adjustable speed drives and arc furnaces, etc. distort voltage/current waveforms in non-conformity to their desired form, bringing challenges to maintain the power quality and ensure efficiency. By means of a power electronic interface, FACTS equipments, including APFs, STATCOMs, Dynamic Voltage Restorers (DVRs), Unified Power Flow Controllers (UPFCs) and etc, are required at all levels from transmission to distribution systems and provide quick control over the reactive power, enhancing the grid stability and transmission capability. They connect to the power system in shunt, or series or a combination of both, as shown in Fig. 2.8, which gives the basic working principle of the FACTS devices. The traditional two-level VSC, FCC, NPC, star-connected CHB are compared in [SG02] for FACTS applications. The star- and delta-connected CHB based STATCOM are presented in [AIY07] and [HMA12] to be intended for installation on medium and high voltage industrial and utility power systems. The shunt APF application is pro-

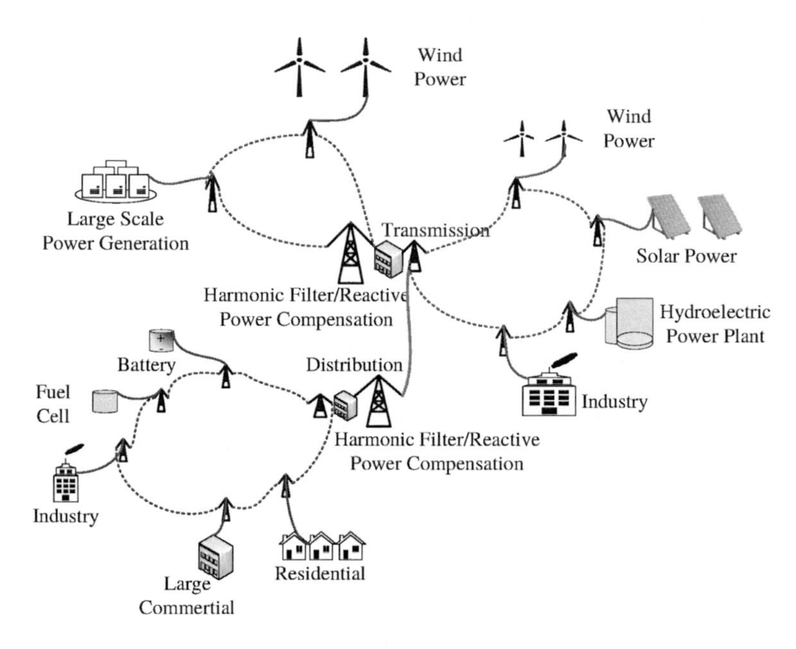

Figure 2.7.: An example of a modern grid

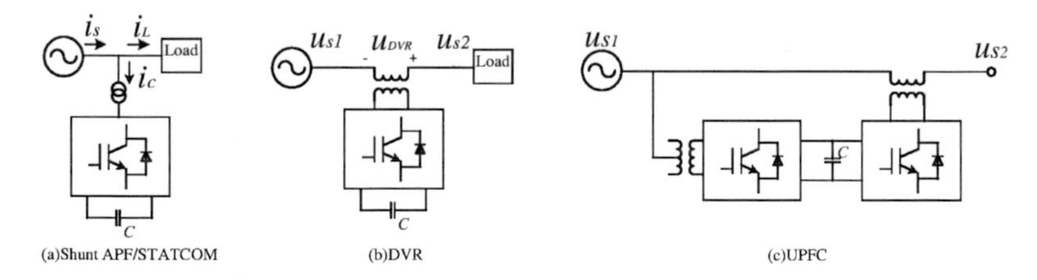

Figure 2.8.: Basic circuits of standardized FACTS devices

posed in [MnCGPA16, AJKM16, WSDC15, YSL^+17, SLZ^+16] respectively with the NPC, FCC, delta-, star-connected CHB and MMC.

2.1.3. Choice of Delta-connected CHB as Shunt APF

Since the power systems are increasingly subject to the requirement to support demanding power-quality applications, a versatile harmonics and reactive power compensation is of increasing importance to provide optimal solutions in order to ensure compliance with increasingly demanding grid codes, thus enabling, for example, the connection of renewable power generation, as well as increase the transmission capacity of existing power lines, thus reducing utilities' investment requirements. The choice of delta-connected CHB is part of the full system solution optimization and takes into account performances, total investment and the operational cost of the system. The delta-connected CHB has the capability to control negative-sequence reactive power because it has the circulating current that flows inside. Due to the superior characteristics and competitive costs of delta-connected CHB multilevel converter, it has been found widespread in practical applications to support the power quality, such as that SVC PLUS products from Siemens and SVC Light products from ABB are based on this converter topology. In some cases, a small high-pass filter is required to support grid code compliance. German network operator Amprion has already installed an SVC PLUS system in the Greater Frankfurt Area. Located in Kriftel in the Rhine-Main region between Frankfurt and Wiesbaden, the unit is one of a number of dynamic compensation systems being used to support the Amprion grid. This configuration confers a number of advantages that explain why it has become so popular:

- independent switching of the converter submodules makes it possible to build up the submodule output voltage in small steps that sum to give a high-quality converter branch voltage waveform - with greatly reduced harmonic content compared to traditional two-level converters
- reduced overall footprint due to reduced need for passive components in the substation yard
- possibility to achieve system redundancy and robustness to failures by series

connection of redundant submodules, reaching extremely high reliability

- possible direct connection to the medium voltage networks without a transformer by series connection of submodules

2.2. Non-ideal Power Supplies

An essential feature of electric power is voltage disturbances [Bon85]. The power system research community has reacted in two ways to the disturbances: disconnection of the disturbance origins to power systems as long as they are detected; and continuous operation with some disturbance mitigation technology since reduced stop times and increased availability result in cost reduction and increased productivity. After the introduction of voltage disturbances in Section 2.2.1, a mathematical representation of periodic voltages with harmonics and unbalance is shown in Section 2.2.2.

2.2.1. Voltage Disturbances in Electric Power Systems

Various voltage disturbances concerning the frequency, amplitude, waveform and symmetry appear as following arts, which can also been shown as Fig. 2.9.

Rapid Voltage Fluctuation (Responsible for Flicker)

This case depends on two factors, the amplitude and frequencies of the voltage fluctuation waveform. This phenomenon is mainly produced by the loads with rapid variable operational modes such as rolling mills, arc furnace, welding machine etc., and can appear in two main forms:

- permanent periodic or erratic variations in the 0.05 Hz to 35 Hz band (arc furnace) with particular sensitivity of the human eye to 10 Hz [Sch99]
- step voltage changes (welding machines)

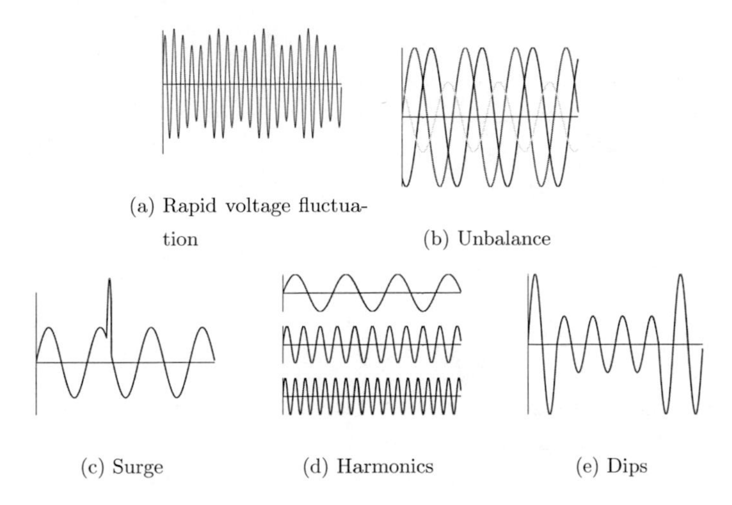

(a) Rapid voltage fluctuation

(b) Unbalance

(c) Surge

(d) Harmonics

(e) Dips

Figure 2.9.: Various arts of voltage disturbance

Unbalance

Electrically dissymmetrical loads connected to three-phase networks give rise to voltage unbalance owing to the fact that the currents absorbed in the phases are not identical. It may be caused by large and/or unequal distribution of single-phase load such as single-phase ac traction, phase to phase loads such as a single-phase welder at phase to phase voltage and unbalanced three-phase loads, for example, when the three-phase loads are comprised of both single- and three-phase equipments. Unequal impedances of a three-phase transmission and distribution power system can also result in voltage unbalance. The unbalance causes nagetive effects. For instance, when three-phase motors are fed by unbalanced voltages, which cause additional negative-sequence currents to circulate in the motors, increasing the heat losses primarily in the rotor. It is illustrated in [IEE94] when the phase-voltage unbalance exceeds 2%, the motor is likely to become overheated if it is operating close to full load.

Surges

Surges are abrupt, short duration increases in voltages. It is in question actually an energy spike. They can be caused by an electromagnetic pulse, electrostatic discharge, the switching of equipment as well as power supply systems, occasionally by lightning but most hazardous. There are several surge protection methods. For example, a type of surge protector attempts to limit the voltage supplied to an electric device by shorting to ground any unwanted voltages above a safe threshold.

Harmonics

The frequency range corresponding to harmonics generally falls between 100 Hz and 2000 Hz inclusive, i.e. between the 2nd and 40th harmonics. They are mainly caused by [CHX+04] device nonlinearity, non-sinusoidal flux distribution in the stator of electric machines, periodically switching of power electronic equipment such as electronic converters.

Voltage Dips and Short Voltage Interruptions

Voltage signals from 10% to 90% of the RMS nominal voltage for 0.5 cycle to 1 min are considered as voltage dips [AR99] according to IEEE stansard p1346. Short voltage interruptions mean the amplitude of the voltage dips attains 100%. Voltages dips are caused by faults in the grid (short circuits), motor starting, fast reclosing of circuit breakers, transformers energizing, etc. The majorities of this fault are not permanent and can be cleared by a simple disconnection, or the faulty part of the network. A voltage dip can be represented with almost instantaneous appearance of low order harmonics and fast increase of their values.

2.2.2. General Description of Periodic Voltages in abc Frame

Here the method of symmetrical components postulates that a three-phase unbalanced system of voltages (of fundamental and harmonic frequency) may be presented by the following three separate components: positive-, negative- and zero-sequence. Imbalance caused by zero-sequence components can only appear in three-phase, four-wire systems. No matter if there exists the neutral line the positive- and negative-sequence components can be present.

- Positive sequence system (a balanced three-phase system in normal sequence)
- Negative sequence system (a balanced three-phase system in reversed sequence)
- Zero sequence system (three phases equal in magnitude and phase angle revolving in positive sequence)

The fundamental and harmonic orders in three-phase generic voltages and currents can be separated and decomposed into symmetrical components. Fig. 2.10 gives an example of the n-th order of a real voltage system composed of positive-, negative- and zero-sequence components. The voltages which contain harmonics and unbalance can be written as Eq. (2.1), with the help of the method of symmetrical components.

$$\begin{cases} v_{sa} = \sum_{n=1}^{\infty} v_{san} = & \sum_{n=1}^{\infty} v_{sa+n} + \sum_{n=1}^{\infty} v_{sa-n} + \sum_{n=1}^{\infty} v_{sa0n} \\ = & \sum_{n=1}^{\infty} \sqrt{2} V_{s+n} \sin(n\omega t + \phi_{vs+n}) \\ + & \sum_{n=1}^{\infty} \sqrt{2} V_{s-n} \sin(n\omega t + \phi_{vs-n}) \\ + & \sum_{n=1}^{\infty} \sqrt{2} V_{s0n} \sin(n\omega t + \phi_{vs0n}) \\ v_{sb} = \sum_{n=1}^{\infty} v_{sbn} = & \sum_{n=1}^{\infty} v_{sb+n} + \sum_{n=1}^{\infty} v_{sb-n} + \sum_{n=1}^{\infty} v_{sb0n} \\ = & \sum_{n=1}^{\infty} \sqrt{2} V_{s+n} \sin\left(n\omega t + \phi_{vs+n} - \frac{2\pi}{3}\right) \\ + & \sum_{n=1}^{\infty} \sqrt{2} V_{s-n} \sin\left(n\omega t + \phi_{vs-n} + \frac{2\pi}{3}\right) \\ + & \sum_{n=1}^{\infty} \sqrt{2} V_{s0n} \sin(n\omega t + \phi_{vs0n}) \\ v_{sc} = \sum_{n=1}^{\infty} v_{scn} = & \sum_{n=1}^{\infty} v_{sc+n} + \sum_{n=1}^{\infty} v_{sc-n} + \sum_{n=1}^{\infty} v_{sc0n} \\ = & \sum_{n=1}^{\infty} \sqrt{2} V_{s+n} \sin\left(n\omega t + \phi_{vs+n} + \frac{2\pi}{3}\right) \\ + & \sum_{n=1}^{\infty} \sqrt{2} V_{s-n} \sin\left(n\omega t + \phi_{vs-n} - \frac{2\pi}{3}\right) \\ + & \sum_{n=1}^{\infty} \sqrt{2} V_{s0n} \sin(n\omega t + \phi_{vs0n}) \end{cases} \quad (2.1)$$

where ω is the fundamental angular frequency, n is the frequency order, subscript $+$, $-$ and 0 represents the positive-, negative- and zero-sequence component; V_{s+n}, V_{s-n}, V_{s0n} and ϕ_{vs+n}, ϕ_{vs-n}, ϕ_{vs0n} are the RMS values and phase angles of the corresponding sequence component of the n-th order voltages.

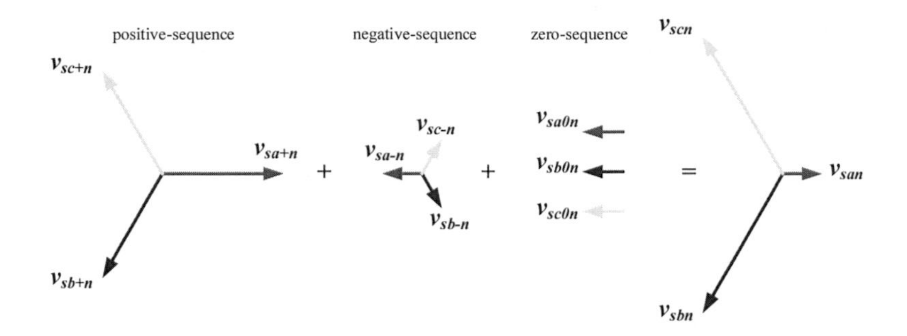

Figure 2.10.: The positive-, negative- and zero-sequence component of an unbalanced three-phase voltage system of the n-th frequency order

2.3. Challenges of Operational Issues of Delta-connected CHB as Shunt APF under Non-ideal Power Supplies

Optimal Current Strategy The difficulties in the optimal operating current strategy for the delta-connected CHB-based shunt APF to minimize the operational power while satisfying the constraints lie in

1. optimization problem formulation
2. to find the optimal solution for the formulated problem

The optimization problem formulation (also called mathematical program) represents a problem choice of decision variables and seeks values that maximize (or minimize) the objective function of the decision variables subject to constraints on variable values expressing the limits on possible decision choices. It is necessary first to formulate the problem in a manner not only reflecting the situation being modelled, but so as to be amenable to computational techniques. Constraints can be expressed in more than one way, and some forms of expression may be more convenient in one context than another. A good problem formulation targets for the existence of a unique solution is assured, and the solution is stable in the sense of being affected only slightly when the data elements of the problem are shifted slightly.

In order to find the optimal solution for the formulated optimization problem, then it heavily depends on the problem convex or non-convex. For the former, there is a huge theory and variety of methods. If the problem is non-convex, it is not simple to solve it. There are two families in the field of solving the optimization problems. One is stochastic methods, such as genetic algorithm, which solves smooth or non-smooth optimization problems with any types of constraints. These methods are easy to apply, however, they cannot guarantee the convergence to an optimal solution in finite time. Another is the deterministic methods which can converge to the optimal solution in finite time, but they generally require that the functions have some properties like being twice continuously differentiable or else.

This dissertation tries to formulate the problem in a way that it can be classified into a specific optimization programming category, and to solve the optimization problem with deterministic methods due to that they are mathematically sound.

Harmonic Interaction Analysis Power supply quality is now a major issue worldwide making harmonic analysis an essential element in power system planning, designing and operation. Power system harmonic interaction analysis presents novel analytical and modelling tools for the assessment of components and systems, and their interactions at harmonic frequencies. The recent proliferation of power electronic equipment is a significant source of harmonic distortion and the dissertation presents effective techniques to tackle this real engineering problem. The challenges of harmonic interaction analysis for the delta-connected CHB-based shunt APF lie in

1. decoupling of the symmetrical components while preserving the coupling with other variables

2. establishing a systematic analysis method

The harmonic interaction analysis method in present references for three-phase systems often deals with the individual phase separately. They are capable of decoupling harmonics in each phase. However, these methods did not treat the three-phase system as an integrity, thus the decoupling of symmetrical components, which is significant to analyse unbalanced systems, has not been achieved.

Further, it is important that the harmonic interaction analysis approach can be easily extended to a high number of harmonics so that it can be applied to a larger network which contains higher frequency orders. This aspect is important that the approach is applicable for other power-electronic and power systems.

This dissertation tries to build a systematic harmonic interaction analysis method, in which the symmetrical components in each frequency order in the variables can be decoupled while preserving the coupling with variables that also contain harmonic sequences, filling in the gap in the conventional studies, providing a great tool for the understanding, designing and operation of the delta-connected CHB as shunt APF.

3. Optimal Current Operation Strategy

In this chapter, a novel optimization-based shunt APF operation strategy is proposed that not only determines the desired APF terminal currents, but also provides the desired circulating current to achieve the minimum operational power of delta-connected CHB-based APF to meet the recommended practice for power quality and other constraints. The strategy is designed to work with non-ideal grid conditions and makes use of a mathematical quadratic optimization with non-convex quadratic constraints. In order to solve this non-convex problem efficiently, the sequential convex programming is applied via an iteration of solving convex subproblems which are achieved by proper linearization of the nonconvex components in the constraints.

3.1. Definitions for Electric Quantity Measurement

Voltage Unbalance

Voltage unbalance in the three-phase electric system is a condition in which the three-phase voltage differs in amplitude and/or does not have its normal 120 degree phase relationship. The most popular voltage unbalance indicator is the ratio of the negative- to positive-sequence component, which is determined using the RMS values at the corresponding frequency, and is known as the negative-sequence Voltage Unbalance Factor (VUF) [IEE94].

$$negative - sequence\ VUF = \frac{RMS\ of\ negative - sequence\ voltage}{RMS\ of\ positive - sequence\ voltage} \quad (3.1)$$

Similarly the ratio of the zero- to positive-sequence voltage component represents the zero-sequence VUF,

$$zero - sequence\ VUF = \frac{RMS\ of\ zero - sequence\ voltage}{RMS\ of\ positive - sequence\ voltage} \tag{3.2}$$

Voltage Distortion

Interharmonics and Harmonics: Interharmonics are frequency components of a periodic quantity that is not an integer multiple of the frequency at which the supply system is operating, for example 50 Hz or 60 Hz. Harmonics are components of integer multiple order greater than one. For example, in a 60 Hz system, the harmonic order 3, also known as the third harmonic, is 180 Hz.

Total Demand Distortion (TDD): The ratio of the RMS of the harmonic content to that of the maximum demand current, considering harmonic components up to the 50th order and specifically excluding interharmonics. Harmonic components of order greater than 50 may be included when necessary.

Total Harmonic Distortion (THD): The ratio of the RMS of the harmonic content to that of the fundamental component, considering harmonic components up to the 50th order and specifically excluding interharmonics. Harmonic components of order greater than 50 may be included when necessary.

According to IEEE Standard 1459, for steady-state conditions, a non-sinusoidal periodical instantaneous voltage as Eq. (2.1) has two distinct components: the power system frequency components v_{sj1} and the remaining terms v_{sjH}, where $j = a, b, c$ represents three phases respectively,

$$v_{sj} = v_{sj1} + v_{sjH} \tag{3.3}$$

Single-phase Voltage THD The total harmonic distortion of the phase-j voltage can be calculated as the following,

$$THD_{v_{sj}} = \frac{V_{sjH}}{V_{sj1}} \tag{3.4}$$

$$V_{sjH} = \sqrt{\sum_{n=2}^{\infty} V_{sjn}^2} \quad (j = a, b, c) \tag{3.5}$$

in Eq. (3.4) and (3.5) V_{sjn} is the effective value of the nth ($n = 1, 2, 3, \cdots$) frequency order voltage of phase-j.

Three-phase Voltage THD THD of a three-phase voltage system can be calculated as Eq. (3.6),

$$THD_{v_s} = \frac{V_{seH}}{V_{se1}} \tag{3.6}$$

$$V_{seH} = \sqrt{\frac{\sum_{n=2}^{\infty}(V_{san}^2 + V_{sbn}^2 + V_{scn}^2)}{3}}, \quad V_{se1} = \sqrt{\frac{V_{sa1}^2 + V_{sb1}^2 + V_{sc1}^2}{3}} \tag{3.7}$$

Power Factor of Three-phase Systems

The effective three-phase voltage V_{se} can be written as

$$V_{se} = \sqrt{V_{se1}^2 + V_{seH}^2} \tag{3.8}$$

with V_{se1} and V_{seH} same as Eq. (3.6). The effective apparent power from the source can be calculated as

$$S_{se} = 3V_{se}I_{se} \tag{3.9}$$

where I_{se} is the effective value of the three-phase currents. The average active power of three-phase systems is given as Eq. (3.10),

$$p_{s,dc} = \sum_{j}\sum_{n=1}^{\infty} V_{sjn}I_{sjn}\cos(\phi_{vsjn} - \phi_{isjn}) \tag{3.10}$$

in which $\phi_{vsjn} - \phi_{isjn}$ is the phase angle deviation between the voltage v_{sjn} and the current i_{sjn}. The power factor of three-phase systems can be calculated as

$$pf = \frac{p_{s,dc}}{S_{se}} \tag{3.11}$$

3.2. Recommended Practice for Electric Power Systems

Electrical energy is a product and, like any other product, should satisfy the proper quality requirements. At the same time, However, the equipment used today often causes voltages/currents distortion in the installation, because of its nonlinear characteristics. Thus, maintaining satisfactory power quality is a joint responsibility for the supplier and the electricity user. Here some of the standards concerning the voltage/current characteristics are reviewed.

Voltage/Current Unbalance Limits

IEEE STD 1159 The recommended practice IEEE STD 1159 [IEE09] describes nominal conditions and deviations from these nominal conditions that may originate within the source of supply or load equipment or may originate from interactions between the source and the load. IEEE STD 1159 mentions that at steady state the current unbalance factor up to 30% and the voltage unbalance factor up to 2% are the typical characteristics of power system electromagnetic phenomena.

EN 50160 According to EN 50160[EuS], which offers the main voltage parameters and their allowable deviation ranges at PCC of low- and medium-voltage power supply, the negative-sequence voltage unbalance factor of the fundamental frequency should be up to 2% for 95% of week, in some locations up to 3%, where the RMS values are evaluated over a fixed interval of 10 min over the observation period of one week, including the weekend. This is a European standard but it is supplemented in some regions or countries by other supplemental standards.

Australia National Electricity Rules Current balance requirements are also set in different locations, such as in Australia National Electricity Rules,

- for connections at voltages less than 30 kV, the current in any phase is not greater than 105% or less than 95% of the average of the currents in the three phases
- for connections at 30 kV or higher voltage, the deviation range of current in any phase is not greater than ±2% of the average of the currents in the three phases

Voltage/Current Distortion Limits

IEEE STD 519 IEEE STD 519-2014 [IEE14] addresses the steady state limitations of the voltage and current waveform distortion. IEEE STD 519-2014 lists the current harmonic distortion limits of systems rated from 120 V to 69 kV, rated 69 kV through 161 kV and above 161 kV. It should be noticed that under transient conditions exceeding these limitations may be encountered. This document sets the quality of power that is to be provided at PCC.

IEC 1000-3-6 This technical report 1000-3-6 deals with the assessment of emission limits for distorting and fluctuating loads in medium-voltage power systems. There is no significant difference between IEC 1000-3-6 and IEEE STD 519 for harmonic voltage distortion limits for medium-voltage networks. But the current limits in IEC 1000-3-6 are added and estimated for the purpose of ensuring voltage distortion limits to be satisfied. The IEC standard makes the current limits more case and system dependent, which is supposed to result in fewer restrictions to customers. However, the calculation of current limits relies on many assumptions, which could defeat the good intentions of the IEC standard [Xu00].

It should be noticed that many transmission system and distribution system operators and national organizations have their own limits especially for high power equipment connecting to medium and high voltage grids. In some circumstances, it is not necessarily that these standards must be satisfied, for instance, when the limits exceed those in EN 50160, many suppliers do not undertake the responsibility since they expound EN 50160 as principally informative. However, even meeting the demands in EN 50160 does not make sure sufficient active and reactive power for some consumers. To solve such conflict the level of active and reactive power required must be negotiated between suppliers and consumers.

3.3. Operation Principle of Delta-connected CHB

The topology of delta-connected CHB presented in this dissertation is shown in Fig. 3.1. The power supply $v_{ga,gb,gc}$ which has associated inductor L_s and resistor R_s feeds the nonlinear load. Here the impedances of the three-phase power supply are assumed equal. At the point of common coupling (PCC), the nonlinear load is

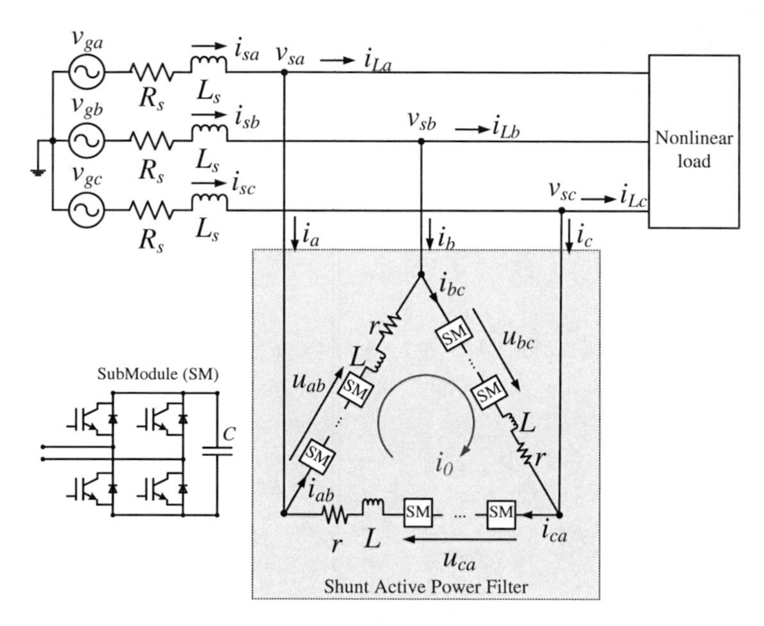

Figure 3.1.: A delta-connected CHB as shunt APF connected to the line

connected in parallel to the delta-connected CHB. Each branch of the CHB consists of cascaded connection of N submodules and each submodule (SM) capacitor has the same capacitance C. The resistor and inductor of three branches of the delta-connected CHB are considered to have the same values, which are represented by r and L respectively. In Fig. 3.1 $u_{ab,bc,bc}$ is the branch voltage produced by the series of submodules in the corresponding branch, $i_{ab,bc,bc}$ are the branch currents. Here $v_{sa,sb,sc}$ is the PCC voltage. The delta configuration of the three branches makes a circulating current i_0 flows inside the delta-connected branches possible.

The SM capacitor voltage sum in branch-ab, bc, ca are

$$u_{Cab} = \sum_{i=1}^{N} u_{Cab}^{\mathrm{SM}_i}, \quad u_{Cbc} = \sum_{i=1}^{N} u_{Cbc}^{\mathrm{SM}_i}, \quad u_{Cab} = \sum_{i=1}^{N} u_{Cca}^{\mathrm{SM}_i} \tag{3.12}$$

By defining $C_{sum} = C/N$, the SM capacitor energy sum in branch-ab, bc, ca are

$$w_{ab} = \frac{1}{2} C_{sum} u_{Cab}^2, \quad w_{bc} = \frac{1}{2} C_{sum} u_{Cbc}^2, \quad w_{ca} = \frac{1}{2} C_{sum} u_{Cca}^2 \tag{3.13}$$

The relationship of the branch current i_{ij}, the terminal current i_j (also as compensating currents), and the circulating current i_0 is

$$\begin{cases} i_{ab} = \frac{i_a - i_b}{3} + i_0 \\ i_{bc} = \frac{i_b - i_c}{3} + i_0 \\ i_{ca} = \frac{i_c - i_a}{3} + i_0 \end{cases} \tag{3.14}$$

Regarding the operational and technological issues of delta-connected CHB, most of the challenges are concerned with some key phenomena that are specific to this converter and provide opportunities for unique design and control optimization. Among the key phenomena the following should be paid attention:

Circulating Current The delta configuration allows negative-sequence reactive power compensation by letting a zero-sequence current circulate inside. Consider each branch as a voltage source, the circulating current flows in the delta configuration in such a manner that it discharges the voltage source having the greater voltage and charges the voltage source having the smaller voltage so that the three voltage sources become equal. Without circulating current it would lead to that one branch of CHB absorbs active power from grid while others release active power to grid, as a result of the dc-link capacitor voltages being out of control. After adding zero-sequence circulating current, dc-link capacitor voltages can be kept stable, when the average active power released and absorbed by per-phase branch is balanced. The instantaneous active power of the three branches are described as

$$\begin{cases} p_{ab} = v_{sab} i_{ab} \\ p_{bc} = v_{sbc} i_{bc} \\ p_{ca} = v_{sca} i_{ca} \end{cases} \tag{3.15}$$

in which the PCC line-line voltages are

$$v_{sab} = v_{sa} - v_{sb}, \quad v_{sbc} = v_{sb} - v_{sc}, \quad v_{sca} = v_{sc} - v_{sa} \tag{3.16}$$

At steady state, the dc component of p_{ab}, p_{bc}, and p_{ca} seen in Eq. (3.15) should be zero in oder to achieve submodule capacitor voltages not drifting. Putting

Eq. (3.14) into Eq. (3.15), Eq (3.17) should be satisfied.

$$\begin{cases} p^*_{ab,dc} = \left(v_{sab}\frac{i^*_a - i^*_b}{3}\right)_{dc} + (v_{sab}i^*_0)_{dc} = 0 \\ p^*_{bc,dc} = \left(v_{sbc}\frac{i^*_b - i^*_c}{3}\right)_{dc} + (v_{sbc}i^*_0)_{dc} = 0 \\ p^*_{ca,dc} = \left(v_{sca}\frac{i^*_c - i^*_a}{3}\right)_{dc} + (v_{sca}i^*_0)_{dc} = 0 \end{cases} \tag{3.17}$$

where $*$ indicates reference value, subscript dc represents the dc component. The sum of the three equations in Eq. (3.17) is equal to zero, indicating that the zero-sequence circulating current just causes the redistribution of average active power among three branches without any influence on the total value of average active power. The redistributed power could be used for canceling that is caused by unbalanced compensating current, as well as for providing proper amount of power for balancing of SM capacitor sum voltages among three branches. More details about how to calculate circulating current can be seen in [WL17].

Submodule Capacitor Voltage Dynamics When the capacitors are as energy storage elements in CHB submodules, the ripples with several order harmonics of the line frequency are unavoidably present on the submodule capacitor voltages. The presence of these harmonics is directly impacting capacitance size and the controller design. It is typical for a converter with CHB to choose the capacitance to limit the capacitor voltage ripple up to $\pm 5\%$ to the nominal dc capacitor voltage to maintain sufficient capacitor voltages for effective current control and increasing the life expectancy of the capacitors. Currently, the capacitor voltage harmonics are ignored and only the average capacitor voltages are taken into consideration in most of the literatures. The research regarding the capacitor voltage ripple can be mainly categorized into two aspects:

1. capacitor voltage ripple suppression by means of different circulating current injection

2. successful operation of the converters by allowing large capacitor voltage variations, which implies the use of significantly reduced capacitance

More details about capacitor voltage dynamics research from the second aspect can be seen in the next chapter.

3.4. The Proposed Strategy for Desired Terminal Currents and Circulating Current Calculation

Aiming at the minimum APF power operation, which is equivalent to minimum APF effective branch currents, the proposed APF operation strategy must additionally guarantee that the desired grid currents satisfy the IEEE-STD 519 current harmonic limitation and IEEE-STD 1159 current unbalance characteristics, a minimum power factor, as well as the internal and external power balance requirement, respectively. In this optimal strategy, there are following constraints:

1. the average active power delivered by the source equals to that consumed by the load
2. the average active power balance among the branches
3. lower bound on the power factor
4. total harmonic distortion limits for source current in each phase
5. individual harmonic distortion limits for source current in each phase
6. negative-sequence unbalance factor limits for source current of the fundamental-frequency component

In order to meet these requirements, the measured PCC voltages ($\boldsymbol{v}_s^{abc}$) and load currents ($\boldsymbol{i}_L^{abc}$) are pre-processed in a harmonic-based manner, before an optimization problem can be formulated. The block diagram for the generation of optimal operation strategy is shown in Fig. 3.2. The diagram consists of three parts, namely, harmonic-based distortion analysis in *dq* frame, optimization algorithm

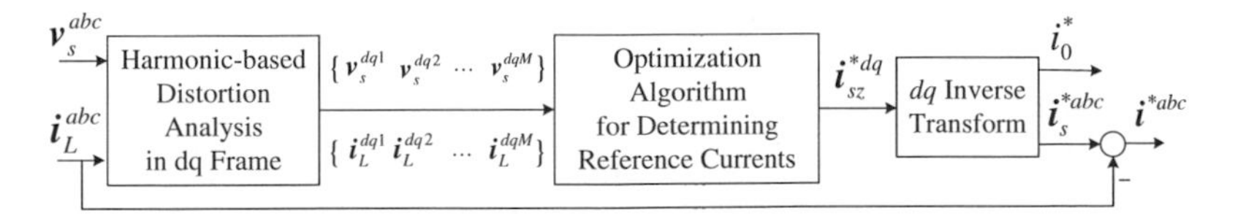

Figure 3.2.: Block diagram of the optimal operation algorithm

for determining reference currents, and dq inverse transform. It is inappropriate to formulate the strategy under the time-varying abc or $\alpha\beta$ frame since the effective branch current calculation concerns the integral of squared instantaneous branch currents in time-varying frames. The strategy is thus formulated in time-invariant dq frame. With the dq values of each frequency component in the measured PCC voltages and load currents, the optimization algorithm calculates the vector $\boldsymbol{i}_{sz}^{*dq}$, which collects the dq values of reference source currents and circulating current. With inverse dq transform, the reference source currents $\boldsymbol{i}_s^{*abc}$ and circulating current $\boldsymbol{i}_0^*$ under the abc frame can be obtained. The reference terminal currents can be calculated by substracting the load currents $\boldsymbol{i}_L^{abc}$ from the obtained source currents $\boldsymbol{i}_s^{*abc}$.

3.4.1. Harmonic-based Distortion Analysis in dq Frame

In general, the measured three-phase PCC voltages v_{sj} and load currents $i_{Lj} (j \in \{a, b, c\})$, seen in Fig. 3.1, can be unbalanced and distorted as mentioned in Chapter 2, as a result they can be summerized in Eq. (3.20) and Eq. (3.21), satisfying a generic description using Eq. (3.18).

$$\boldsymbol{x}^{abcn} = \begin{bmatrix} x_{an} \\ x_{bn} \\ x_{cn} \end{bmatrix} = \underbrace{\begin{bmatrix} x_{a+n} \\ x_{b+n} \\ x_{c+n} \end{bmatrix}}_{\boldsymbol{x}^{abc+n}} + \underbrace{\begin{bmatrix} x_{a-n} \\ x_{b-n} \\ x_{c-n} \end{bmatrix}}_{\boldsymbol{x}^{abc-n}} + \underbrace{\begin{bmatrix} x_{a0n} \\ x_{b0n} \\ x_{c0n} \end{bmatrix}}_{\boldsymbol{x}^{abc0n}} \tag{3.18}$$

$$\boldsymbol{x}^{abc+n} = \underbrace{\sqrt{\frac{2}{3}} \begin{bmatrix} \cos\theta_n & -\sin\theta_n \\ \cos\left(\theta_n - \frac{2\pi}{3}\right) & -\sin\left(\theta_n - \frac{2\pi}{3}\right) \\ \cos\left(\theta_n + \frac{2\pi}{3}\right) & -\sin\left(\theta_n + \frac{2\pi}{3}\right) \end{bmatrix}}_{\boldsymbol{T}_{+n}^{2r/3s}} \underbrace{\begin{bmatrix} x_{d+n} \\ x_{q+n} \end{bmatrix}}_{\boldsymbol{x}^{dq+n}}$$

$$\boldsymbol{x}^{abc-n} = \underbrace{\sqrt{\frac{2}{3}} \begin{bmatrix} \cos\theta_n & -\sin\theta_n \\ \cos\left(\theta_n + \frac{2\pi}{3}\right) & -\sin\left(\theta_n + \frac{2\pi}{3}\right) \\ \cos\left(\theta_n - \frac{2\pi}{3}\right) & -\sin\left(\theta_n - \frac{2\pi}{3}\right) \end{bmatrix}}_{\boldsymbol{T}_{-n}^{2r/3s}} \underbrace{\begin{bmatrix} x_{d-n} \\ x_{q-n} \end{bmatrix}}_{\boldsymbol{x}^{dq-n}}$$

$$\boldsymbol{x}^{abc0n} = \underbrace{\begin{bmatrix} \cos\theta_n & -\sin\theta_n \\ \cos\theta_n & -\sin\theta_n \\ \cos\theta_n & -\sin\theta_n \end{bmatrix}}_{\boldsymbol{T}_{0n}^{2r/3s}} \underbrace{\begin{bmatrix} x_{d0n} \\ x_{q0n} \end{bmatrix}}_{\boldsymbol{x}^{dq0n}}$$

where $\theta_n = n\omega t$ with ω as fundamental line-frequency, the superscript $2r/3s$ in $\boldsymbol{T}_{+n}^{2r/3s}$, $\boldsymbol{T}_{+n}^{2r/3s}$ and $\boldsymbol{T}_{0n}^{2r/3s}$ indicates transferring two-phase rotating signals to three-phase stationary signals. x_{ab0n}, x_{bc0n}, and x_{ca0n} contained in $\boldsymbol{x}^{abc0n}$ exists $x_{ab0n} = x_{bc0n} = x_{ca0n}$. Putting x_{ab0n} at α and $x_{ab0n,shift}$ at β coordinate, where $x_{ab0n,shift}$ is intentionally introduced by shifting the original zero-sequence x_{ab0n} by $\frac{\pi}{2}$,

$$\boldsymbol{x}^{\alpha\beta0n} = \begin{bmatrix} x_{ab0n} & x_{ab0n,shift} \end{bmatrix}^{\top} \tag{3.19}$$

$\boldsymbol{x}^{dq0n}$ in Eq. (3.18) can be calculated

$$\boldsymbol{x}^{dq0n} = \boldsymbol{T}_{0n}^{2s/2r}\boldsymbol{x}^{\alpha\beta0n}, \text{ where } \boldsymbol{T}_{0n}^{2s/2r} = \begin{bmatrix} \cos\theta_n & \sin\theta_n \\ -\sin\theta_n & \cos\theta_n \end{bmatrix}, \text{ with } \boldsymbol{T}_{0n}^{2r/2s} = \left(\boldsymbol{T}_{0n}^{2s/2r}\right)^{-1}$$

x_{ab0n} is then $x_{ab0n} = \begin{bmatrix} \cos\theta_n & -\sin\theta_n \end{bmatrix} \boldsymbol{x}^{dq0n}$, shown in Eq. (3.18).

Eq. (3.18) shows how the variable in *abc*-frame can be expressed by the time-invariant *dq* variables. Eq. (3.20) and Eq. (3.21) are then

$$\boldsymbol{v}_s^{abc} = \sum_{n=1}^{M} \boldsymbol{v}_s^{abcn} = \sum_{n=1}^{M} \left(\boldsymbol{v}_s^{abc+n} + \boldsymbol{v}_s^{abc-n} + \boldsymbol{v}_s^{abc0n}\right) \tag{3.20}$$

$$\boldsymbol{i}_L^{abc} = \sum_{n=1}^{M} \boldsymbol{i}_L^{abcn} = \sum_{n=1}^{M} \left(\boldsymbol{i}_L^{abc+n} + \boldsymbol{i}_L^{abc-n}\right) \tag{3.21}$$

In Eq. (3.20) and Eq. (3.21) n is the harmonic order. Eq. (3.21) is assumed to not contain the zero-sequence component. M is the highest harmonic order under consideration with $M \leq 40$, indicating that only low-frequency orders are taken into account since currently only limits for harmonics up to the 40th order are provided in the standardization, although electrical devices with active power electronic on the ac-side have harmonic emissions in a higher frequency range especially at the switching frequency and multiples of it. High frequencies, whether or not these frequencies are harmonics or interharmonics have little significance. All studies of this work are made in the frequency range up to 40th.

The reference source currents and circulating current to be determined later are expressed as Eq. (3.22) and Eq. (3.23), with $i^*_{sd\pm n}$, $i^*_{sq\pm n}$ as nth order positive and negative sequence dq variables of reference source currents,

$$\boldsymbol{i}_s^{*abc} = \sum_{n=1}^{M} \boldsymbol{i}_s^{*abcn} = \sum_{n=1}^{M} \left(\boldsymbol{i}_s^{*abc+n} + \boldsymbol{i}_s^{*abc-n} \right) \tag{3.22}$$

and i^*_{d0n}, i^*_{q0n} as nth order zero sequence dq variables of reference circulating currents (as i^*_0 only contains zero sequence)

$$i^*_0 = \sum_{n=1}^{M} \left(i^*_{d0n} \sin n\omega t + i^*_{q0n} \cos n\omega t \right) \tag{3.23}$$

A state vector $\boldsymbol{i}_{sz}^{*dq}$ can be defined based on the dq values of $\boldsymbol{i}_s^{*abc}$ and i^*_0,

$$\boldsymbol{i}_{sz}^{*dq} = \begin{bmatrix} \boldsymbol{i}_{sz}^{*dq1\top} & \boldsymbol{i}_{sz}^{*dq2\top} & \cdots & \boldsymbol{i}_{sz}^{*dqM\top} \end{bmatrix}^\top \tag{3.24}$$

$$\boldsymbol{i}_{sz}^{*dqn} = \begin{bmatrix} i^*_{sd+n} & i^*_{sq+n} & i^*_{sd-n} & i^*_{sq-n} & i^*_{d0n} & i^*_{q0n} \end{bmatrix}^\top \tag{3.25}$$

According to Eq. (3.22) and Eq. (3.23) the desired source currents $\boldsymbol{i}_s^{*abc}$ and circulating current i^*_0 can be calculated from the state variable vector $\boldsymbol{i}_{sz}^{*dq}$ which will be used as system variable in the optimization. The APF reference terminal currents of each phase can be derived by substracting the load current from the desired grid current

$$\boldsymbol{i}^{*abc} = \boldsymbol{i}_s^{*abc} - \boldsymbol{i}_L^{abc} \tag{3.26}$$

Although our purpose is to determine $\boldsymbol{i}_{sz}^{*dq}$, the objective in the optimal operation strategy is to minimize the APF apparent power, which is directly related to the branch currents. The desired branch currents, denoted by i^*_{ij}, ($ij \in \{ab, bc, ca\}$), have the following description in abc-frame

$$\begin{bmatrix} i^*_{ab} \\ i^*_{bc} \\ i^*_{ca} \end{bmatrix} = \frac{1}{3} \begin{bmatrix} 1 & -1 & 0 \\ 0 & 1 & -1 \\ -1 & 0 & 1 \end{bmatrix} \begin{bmatrix} i^*_{sa} - i_{La} \\ i^*_{sb} - i_{Lb} \\ i^*_{sc} - i_{Lc} \end{bmatrix} + \begin{bmatrix} 1 \\ 1 \\ 1 \end{bmatrix} i^*_0 \tag{3.27}$$

With the generic description as Eq. (3.18), Eq. (3.21), Eq. (3.22) and Eq. (3.23), the nth order time-invariant dq representation of Eq. (3.27) can be easily derived

$$\boldsymbol{i}^{*dqn} = \boldsymbol{T}_1 \boldsymbol{i}_{sz}^{*dqn} - \boldsymbol{T}_2 \boldsymbol{i}_L^{dqn} \tag{3.28}$$

with $\boldsymbol{i}_{sz}^{*dqn}$ defined in Eq. (3.25), and

$$\begin{aligned} \boldsymbol{i}^{*dqn} &= \begin{bmatrix} i_{d+n}^* & i_{q+n}^* & i_{d-n}^* & i_{q-n}^* & i_{d0n}^* & i_{q0n}^* \end{bmatrix}^\top \\ \boldsymbol{i}_L^{dqn} &= \begin{bmatrix} i_{Ld+n} & i_{Lq+n} & i_{Ld-n} & i_{Lq-n} \end{bmatrix}^\top \end{aligned}$$

where i_{d+n}^*, i_{q+n}^* and i_{d-n}^*, i_{q-n}^* are dq values of the nth order branch currents in positive sequence and negative sequence, and $\boldsymbol{T}_1$ and $\boldsymbol{T}_2$ can be found in Eq. (3.29).

Summarizing all considered frequency orders, with the definition of Eq. (3.24) and

$$\begin{aligned} \boldsymbol{i}^{*dq} &= \begin{bmatrix} \boldsymbol{i}^{*dq1\top} & \boldsymbol{i}^{*dq2\top} & \cdots & \boldsymbol{i}^{*dqM\top} \end{bmatrix}^\top \\ \boldsymbol{i}_L^{dq} &= \begin{bmatrix} \boldsymbol{i}_L^{dq1\top} & \boldsymbol{i}_L^{dq2\top} & \cdots & \boldsymbol{i}_L^{dqM\top} \end{bmatrix}^\top \end{aligned}$$

the time-invariant dq description of Eq. (3.27) is

$$\boldsymbol{i}^{*dq} = \underbrace{(\boldsymbol{I}_M \otimes \boldsymbol{T}_1)}_{\boldsymbol{T}_{sz}} \boldsymbol{i}_{sz}^{*dq} - \underbrace{(\boldsymbol{I}_M \otimes \boldsymbol{T}_2)}_{\boldsymbol{T}_L} \boldsymbol{i}_L^{dq} \tag{3.30}$$

where $\boldsymbol{I}_M$ is identity matrix of the dimension $M \times M$.

$$\boldsymbol{T}_1 = \text{blkdiag}\left(\begin{bmatrix} \frac{1}{2} & -\frac{\sqrt{3}}{6} \\ \frac{\sqrt{3}}{6} & \frac{1}{2} \end{bmatrix}, \begin{bmatrix} \frac{1}{2} & \frac{\sqrt{3}}{6} \\ -\frac{\sqrt{3}}{6} & \frac{1}{2} \end{bmatrix}, \begin{bmatrix} 1 & 0 \\ 0 & 1 \end{bmatrix} \right), \quad \boldsymbol{T}_2 = \boldsymbol{T}_1(:, 1:4) \tag{3.29}$$

where $\boldsymbol{T}_1(:, 1:4)$ represents is the column 1 to 4 of $\boldsymbol{T}_1$.

3.4.2. Optimization Problem Formulation

Objective Function

The minimization of the APF apparent power is equivalent to the minimization of effective branch current (I_e) with the known source voltages,

$$I_e = \sqrt{\sum_{n=1}^{M} I_{en}^2} = \sqrt{\sum_{n=1}^{M} \boldsymbol{i}^{*dqn\top}\boldsymbol{Q}_e\boldsymbol{i}^{*dqn}} = \sqrt{\boldsymbol{i}^{*dq\top}\left(\boldsymbol{I}_M \otimes \boldsymbol{Q}_e\right)\boldsymbol{i}^{*dq}} \tag{3.31}$$

where I_{en} is nth effective desired branch currents. Apply Eq. (3.30) to Eq. (3.31),

$$I_e = \sqrt{\boldsymbol{i}_{sz}^{*dq\top}\boldsymbol{Q}_{obj}\boldsymbol{i}_{sz}^{*dq} + \boldsymbol{F}_{obj}\boldsymbol{i}_{sz}^{*dq} + cst} \tag{3.32}$$

where

$$\begin{aligned}
\boldsymbol{Q}_{obj} &= \boldsymbol{T}_{sz}^{\top}(\boldsymbol{I}_M \otimes \boldsymbol{Q}_e)\boldsymbol{T}_{sz} \\
\boldsymbol{F}_{obj} &= -2\boldsymbol{I}_L^{dq\top}\boldsymbol{T}_L^{\top}(\boldsymbol{I}_M \otimes \boldsymbol{Q}_e)\boldsymbol{T}_{sz} \\
cst &= \boldsymbol{I}_L^{dq\top}\boldsymbol{T}_L^{\top}(\boldsymbol{I}_M \otimes \boldsymbol{Q}_e)\boldsymbol{T}_L\boldsymbol{I}_L^{dq} \\
\boldsymbol{Q}_e &= \mathrm{blkdiag}(\frac{1}{3},\frac{1}{3},\frac{1}{3},\frac{1}{3},\frac{1}{2},\frac{1}{2})
\end{aligned}$$

As cst depends only on the measurement $\boldsymbol{i}_L^{dq}$. To avoid the square root calculation, the relevant objective function can be chosen as Eq (3.33), in which $\boldsymbol{i}_{sz}^{*dq}$ is the variable to be determined in the strategy,

$$f_{obj} = \boldsymbol{i}_{sz}^{*dq\top}\boldsymbol{Q}_{obj}\boldsymbol{i}_{sz}^{*dq} + \boldsymbol{F}_{obj}\boldsymbol{i}_{sz}^{*dq} \tag{3.33}$$

Constraints on Average Active Power Balance between Source and Load

The average active power generated by the source ($p_{s,dc}^*$) must be equal to that consumed by the load ($p_{L,dc}$),

$$p_{s,dc}^* = p_{L,dc} \tag{3.34}$$

Thus, the average power constraint is then

$$p_{s,dc}^* = \sum_{n=1}^{M} \boldsymbol{A}_{pn}\boldsymbol{i}_{sz}^{*dqn} = \boldsymbol{A}_p\boldsymbol{i}_{sz}^{*dq} = p_{L,dc} \tag{3.35}$$

with

$$\boldsymbol{A}_{pn} = \begin{bmatrix} V_{sd+n} & V_{sq+n} & V_{sd-n} & V_{sq-n} & 0 & 0 \end{bmatrix}$$
$$\boldsymbol{A}_{p} = \begin{bmatrix} \boldsymbol{A}_{p1} & \boldsymbol{A}_{p2} & \cdots & \boldsymbol{A}_{pM} \end{bmatrix}$$

Constraints on Average Active Power Balance among the Branches

The constraint Eq. (3.34) illustrates that at the steady state the average active power provided by the shunt APF is zero, i.e., $p^*_{ab,dc} + p^*_{bc,dc} + p^*_{ca,dc} = 0$, to make the delta-connected CHB work, the average active power in the CHB branches must be balanced,

$$p^*_{ab,dc} - p^*_{bc,dc} = 0, \quad p^*_{bc,dc} - p^*_{ca,dc} = 0 \tag{3.36}$$

where $p^*_{ij,dc} = (v_{sij} i^*_{ij})_{dc}$, in which v_{sij} is the line-line voltages. The line-line voltages are also with harmonics and unbalance, as the PCC voltages are distorted and unbalanced. No matter if PCC voltages contain zero-sequence components or not, the zero-sequence components in the line-line voltages are zero.

The average active power analysis, the details of which are shown in Section A.1, leads to Eq. (3.37)

$$\begin{bmatrix} p^*_{ab,dc} - p^*_{bc,dc} \\ p^*_{bc,dc} - p^*_{ca,dc} \end{bmatrix} = \sum_{n=1}^{M} \begin{bmatrix} \boldsymbol{A}_{pabbcn} \\ \boldsymbol{A}_{pbccan} \end{bmatrix} \boldsymbol{i}^{*dqn} = \underbrace{\begin{bmatrix} \boldsymbol{A}_{pabbc1} & \boldsymbol{A}_{pabbc2} & \cdots & \boldsymbol{A}_{pabbcM} \\ \boldsymbol{A}_{pbcca1} & \boldsymbol{A}_{pbcca2} & \cdots & \boldsymbol{A}_{pbccaM} \end{bmatrix}}_{\boldsymbol{A}_{\Delta p}} \boldsymbol{i}^{*dqn} \tag{3.37}$$

where $\boldsymbol{A}_{pabbcn}$ and $\boldsymbol{A}_{pbccan}$ are matrices with dimension of 1×6.

$$\boldsymbol{A}_{pabbcn}(1,1) = -\frac{1}{2\sqrt{3}} V_{sllq-n} + \frac{1}{2} V_{slld-n}$$
$$\boldsymbol{A}_{pabbcn}(1,2) = \frac{1}{2\sqrt{3}} V_{slld-n} + \frac{1}{2} V_{sllq-n}$$
$$\boldsymbol{A}_{pabbcn}(1,3) = \frac{1}{2\sqrt{3}} V_{sllq+n} + \frac{1}{2} V_{slld+n}$$
$$\boldsymbol{A}_{pabbcn}(1,4) = -\frac{1}{2\sqrt{3}} V_{slld+n} + \frac{1}{2} V_{sllq+n}$$
$$\boldsymbol{A}_{pabbcn}(1,5) = \frac{1}{2\sqrt{2}} (-V_{sllq+n} + V_{sllq-n}) + \frac{3}{2\sqrt{6}} (V_{slld+n} + V_{slld-n})$$

$$A_{pabbcn}(1,6) = \frac{1}{2\sqrt{2}}(V_{slld+n} - V_{slld-n}) + \frac{3}{2\sqrt{6}}(V_{sllq+n} + V_{sllq-n})$$

$$A_{pbccan} = \Big[\tfrac{1}{\sqrt{3}}V_{sllq-n} \quad -\tfrac{1}{\sqrt{3}}V_{slld-n} \quad -\tfrac{1}{\sqrt{3}}V_{sllq+n} \quad \tfrac{1}{\sqrt{3}}V_{slld+n} \quad \tfrac{1}{\sqrt{2}}(V_{sllq+n} - V_{sllq-n}) \quad \tfrac{1}{\sqrt{2}}(-V_{slld+n} + V_{slld-n})\Big]$$

where V_{slld+n}, V_{sllq+n} and V_{slld-n}, V_{sllq-n} are dq values of nth line-line voltages in positive and negative sequence. There exists following relationship between dq values of the line-line voltages and PCC voltages,

$$\begin{pmatrix} V_{slld+n} \\ V_{sllq+n} \end{pmatrix} = \begin{pmatrix} \frac{3}{2} & -\frac{\sqrt{3}}{2} \\ \frac{\sqrt{3}}{2} & \frac{3}{2} \end{pmatrix} \begin{pmatrix} V_{sd+n} \\ V_{sq+n} \end{pmatrix} \tag{3.38}$$

$$\begin{pmatrix} V_{slld-n} \\ V_{sllq-n} \end{pmatrix} = \begin{pmatrix} \frac{3}{2} & \frac{\sqrt{3}}{2} \\ -\frac{\sqrt{3}}{2} & \frac{3}{2} \end{pmatrix} \begin{pmatrix} V_{sd-n} \\ V_{sq-n} \end{pmatrix} \tag{3.39}$$

Combine Eq. (3.30) to Eq. (3.37), the constraint of Eq. (3.36) becomes

$$(A_{\Delta p} T_{sz})\, i_{sz}^{*dq} - (A_{\Delta p} T_L)\, i_L^{dq} = 0 \tag{3.40}$$

Constraints on Power Factor

The power factor (pf) satisfies

$$pf = \frac{p_{L,dc}}{3V_{se}I_{se}} \geq \underline{pf} \tag{3.41}$$

where V_{se} is the effective value of PCC voltages and can be calculated via $V_{se} = \sqrt{v_s^{dq\top}\left(I_M \otimes Q_e\right) v_s^{dq}}$. I_{se} is the effective value of desired source currents, $\underline{pf}$ is the power factor lower bound. Eq. (3.41) can be rewritten to the following referring to the optimization variable,

$$i_{sz}^{*dq\top} Q_{pf} i_{sz}^{*dq} - p_{L,dc}^2 \leq 0 \tag{3.42}$$

$$Q_{pf} = 9V_{se}^2 \underline{pf}^2 \left(I_M \otimes \text{blkdiag}(\frac{1}{3}, \frac{1}{3}, \frac{1}{3}, \frac{1}{3}, 0, 0) \right)$$

Constraints on Total Harmonic Distortion in Source Currents

The total harmonic distortion of the desired source current in phase-j is defined as,

$$THD_{i^*_{sj}} = \sqrt{\sum_{H=2}^{M} I^2_{sjH}/I_{sj1}} \leq \lambda_j \tag{3.43}$$

in which, I_{sjH} is the RMS value of Hth order source current of phase-j, λ_j is the THD upper bound that satisfies IEEE-519 [IEE14]. Eq. (3.43), rewritten as

$$\sum_{H=2}^{M} I^2_{sjH} - \lambda_j^2 I^2_{sj1} \leq 0 \tag{3.44}$$

equivalent to

$$\boldsymbol{i}_{sz}^{*dq\top} \boldsymbol{Q}_{jH} \boldsymbol{i}_{sz}^{*dq} - \boldsymbol{i}_{sz}^{*dq\top} \boldsymbol{Q}_{j1} \boldsymbol{i}_{sz}^{*dq} \leq 0 \tag{3.45}$$

where

$$\begin{aligned} \boldsymbol{Q}_{jH} &= \text{blkdiag}(\mathbf{0}_6, \quad \boldsymbol{I}_{M-1} \otimes \boldsymbol{Q}_j) \\ \boldsymbol{Q}_{j1} &= \text{blkdiag}(\lambda_j^2 \boldsymbol{Q}_j, \quad \mathbf{0}_{6(M-1)} \quad) \end{aligned}$$

in which $\boldsymbol{Q}_j$ can be found from Eq. (3.49) to Eq. (3.51); $\mathbf{0}_{6(M-1)}$ denotes a squared zero matrix with the dimension $(6(M-1) \times 6(M-1))$.

Constraints on Individual Harmonic Distortion in Source Currents

The individual Hth order source current (IHD) of phase-j is,

$$IHD_{i^*_{sj}} = I_{sjH}/I_{sj1} \leq \lambda_{jH} \tag{3.46}$$

in which, I_{sjH} is the RMS value of Hth order source current of phase-j, λ_{jH} is the IHD upper bound specified by IEEE-519. Eq. (3.46), rewritten as $I^2_{sjH} - \lambda^2_{jH} I^2_{sj1} \leq 0$, is equivalent to

$$\boldsymbol{i}_{sz}^{*dq\top} \boldsymbol{Q}'_{jH} \boldsymbol{i}_{sz}^{*dq} - \boldsymbol{i}_{sz}^{*dq\top} \boldsymbol{Q}'_{j1} \boldsymbol{i}_{sz}^{*dq} \leq 0 \tag{3.47}$$

where

$$\boldsymbol{Q}'_{jH} = \text{blkdiag}\left(\mathbf{0}_{6(H-1)}, \boldsymbol{Q}_j, \mathbf{0}_{6(M-H)}\right), \quad \boldsymbol{Q}'_{j1} = \text{blkdiag}\left(\lambda^2_{jH}\boldsymbol{Q}_j, \mathbf{0}_{6(M-1)}\right) \tag{3.48}$$

with $\boldsymbol{Q}_j$ in Eq. (3.48) is same as Eq. (3.49) to Eq. (3.51); $\mathbf{0}_{6l}$ denotes a squared zero matrix with the dimension $(6l \times 6l)$.

Constraints on Source Current Unbalance of Fundamental Frequency

Quantification of source current unbalance in the fundamental frequency can be achieved by current unbalance factor (CUF) which is defined as the ratio of the RMS of the negative sequence source current (I_{s-1}) to positive sequence component (I_{s+1}). The limitation of unbalance in desired source currents can be specified

---

$$\boldsymbol{Q}_a = \frac{1}{3}\begin{bmatrix} \boldsymbol{I}_2 & \frac{\boldsymbol{I}_2}{2} & \mathbf{0}_2 \\ \frac{\boldsymbol{I}_2}{2} & \boldsymbol{I}_2 & \mathbf{0}_2 \\ \mathbf{0}_2 & \mathbf{0}_2 & \mathbf{0}_2 \end{bmatrix} \tag{3.49}$$

$$\boldsymbol{Q}_b = \frac{1}{3}\begin{bmatrix} 1 & 0 & -\frac{1}{4} & \frac{\sqrt{3}}{4} & 0 & 0 \\ 0 & 1 & -\frac{\sqrt{3}}{4} & -\frac{1}{4} & 0 & 0 \\ -\frac{1}{4} & -\frac{\sqrt{3}}{4} & 1 & 0 & 0 & 0 \\ \frac{\sqrt{3}}{4} & -\frac{1}{4} & 0 & 1 & 0 & 0 \\ 0 & 0 & 0 & 0 & 0 & 0 \\ 0 & 0 & 0 & 0 & 0 & 0 \end{bmatrix} \tag{3.50}$$

$$\boldsymbol{Q}_c = \frac{1}{3}\begin{bmatrix} 1 & 0 & -\frac{1}{4} & -\frac{\sqrt{3}}{4} & 0 & 0 \\ 0 & 1 & \frac{\sqrt{3}}{4} & -\frac{1}{4} & 0 & 0 \\ -\frac{1}{4} & \frac{\sqrt{3}}{4} & 1 & 0 & 0 & 0 \\ -\frac{\sqrt{3}}{4} & -\frac{1}{4} & 0 & 1 & 0 & 0 \\ 0 & 0 & 0 & 0 & 0 & 0 \\ 0 & 0 & 0 & 0 & 0 & 0 \end{bmatrix} \tag{3.51}$$

as

$$CUF = I_{s-1}/I_{s+1} \leq \gamma_1 \tag{3.52}$$

where γ_1 is an upper bound satisfying the current imbalance characteristics in IEEE STD-1159 [IEE09]. Since

$$I_{s-1}^2 = (i_{sd-1}^{*2} + i_{sq-1}^{*2})/3, \quad I_{s+1}^2 = (i_{sd+1}^{*2} + i_{sq+1}^{*2})/3 \tag{3.53}$$

the constraint of Eq. (3.52) can be rewritten as

$$\boldsymbol{i}_{sz}^{*dq\top} \boldsymbol{Q}_{CUF-} \boldsymbol{i}_{sz}^{*dq} - \boldsymbol{i}_{sz}^{*dq\top} \boldsymbol{Q}_{CUF+} \boldsymbol{i}_{sz}^{*dq} \leq 0 \tag{3.54}$$

in which

$$\boldsymbol{Q}_{CUF-} = \text{blkdiag}(0, \quad 0, \quad 1, 1, 0, 0, \boldsymbol{0}_{6(M-1)}) \tag{3.55}$$

$$\boldsymbol{Q}_{CUF+} = \text{blkdiag}(\gamma_1^2, \gamma_1^2, 0, 0, 0, 0, \boldsymbol{0}_{6(M-1)}) \tag{3.56}$$

3.4.3. Combination of Objective Function and Constraints

Combine the objective function Eq. (3.33) with constraints Eq. (3.35), Eq. (3.40), Eq. (3.42), Eq. (3.45), Eq. (3.47) and Eq. (3.54), the following optimization problem is obtained,

$$min \quad f_{obj}(\boldsymbol{i}_{sz}^{*dq}) = \boldsymbol{i}_{sz}^{*dq\top} \boldsymbol{Q}_{obj} \boldsymbol{i}_{sz}^{*dq} + \boldsymbol{F}_{obj} \boldsymbol{i}_{sz}^{*dq} \tag{3.57a}$$

$$s.t. \quad \boldsymbol{A}_p \boldsymbol{i}_{sz}^{*dq} - p_{L,dc} = 0 \tag{3.57b}$$

$$(\boldsymbol{A}_{\Delta p} \boldsymbol{T}_{sz}) \boldsymbol{i}_{sz}^{*dq} - (\boldsymbol{A}_{\Delta p} \boldsymbol{T}_L) \boldsymbol{i}_L^{dq} = \boldsymbol{0} \tag{3.57c}$$

$$\boldsymbol{i}_{sz}^{*dq\top} \boldsymbol{Q}_{pf} \boldsymbol{i}_{sz}^{*dq} - p_{L,dc}^2 \leq 0 \tag{3.57d}$$

$$\boldsymbol{i}_{sz}^{*dq\top} \boldsymbol{Q}_{jH} \boldsymbol{i}_{sz}^{*dq} - \boldsymbol{i}_{sz}^{*dq\top} \boldsymbol{Q}_{j1} \boldsymbol{i}_{sz}^{*dq} \leq 0 \tag{3.57e}$$

$$\boldsymbol{i}_{sz}^{*dq\top} \boldsymbol{Q}'_{jH} \boldsymbol{i}_{sz}^{*dq} - \boldsymbol{i}_{sz}^{*dq\top} \boldsymbol{Q}'_{j1} \boldsymbol{i}_{sz}^{*dq} \leq 0 \tag{3.57f}$$

$$\boldsymbol{i}_{sz}^{*dq\top} \boldsymbol{Q}_{CUF-} \boldsymbol{i}_{sz}^{*dq} - \boldsymbol{i}_{sz}^{*dq\top} \boldsymbol{Q}_{CUF+} \boldsymbol{i}_{sz}^{*dq} \leq 0 \tag{3.57g}$$

Eqs. (3.57a)- (3.57g) belong to a nonconvex QCQP optimization problem, with a convex cost function f_{obj} and nonconvex constraints:

1. The cost function is convex as $\boldsymbol{Q}_{obj}$ is positive definite
2. Eqs. (3.57b)- (3.57c) are linear, Eq. (3.57d) is a convex quadratic constraint
3. Eqs. (3.57e)- (3.57g) are nonconvex due to the difference of two convex functions (DC).

3.4.4. Solution of the Optimization Problem via Sequential Convex Programming

Non-convex problems are difficult to solve, all known algorithms to solve them have a complexity that grows exponentially with problem dimensions. As the non-convex problem is NP-hard, the optimization methods are typically based on convex relaxations of the problem [BST11]. The Semidefinite Programming (SDP) plays useful roles in the formulation of the relaxations for the non-convex problems. By using products of the original variables in the optimization problem as the new variables, the original problem is transformed to the SDP, which is to minimize a linear function with the constraint which is an affine combination of symmetric and positive matrices [VB96]. Lagrangian relaxation, the dual of the SDP relaxation, is also considered as SDP by usage of Schur complement [HKHC10]. It provides a fast solution but may generate numerical convergence [CLL00] and thus the challenging problem is to find an approapriate Karush-Kuhn-Tucker (KKT) solution. Compared with the SDP, in Reformulation-Linearization Technique (RLT) relaxation [VB96, LMS$^+$10, HKHC10], the constraints are strengthened by adding the valid linear inequality constraints on the new variables [HDS98]. SDP and RLT relaxations obtain tighter bounds at the expense of a larger number of variables and constraints, moreover, it needs techniques such as randomization procedure to extract good feasible solution of original nonconvex problem from optimal solution of relaxed problem [Ans09].

A generic algorithm framework for solving nonlinear optimization problems with partially convex structure was proposed which is called Sequential Convex Programming (SCP). In an SCP procedure, a convex approximation of the original non-convex problem is repeatedly solved until convergence. Finding the solution of a non-convex programming problem is thus reduced to solving a collection of convex - hence tractable - programming problems. To the family of SCP methods belong such classical algorithms as the constrained or unconstrained Gauss-Newton methods as well as sequential linear programming or sequential quadratic programming with convex subproblems. All these methods are based on linearization of nonconvex constraints or objective functions, and are widely used in applications of nonlinear optimization.

When non-convex QCQP problems with DC (difference of two convex functions) constraints are treated within an SCP framework, it is possible to only linearise the concave parts. By linearising the concave part at a feasible starting point, the non-convex problem is transformed to a convex problem which can be solved efficiently, leading to a new feasible point with a lower objective value. If we linearise again the non-convex function around the new feasible point and repeat the procedure, we can obtain a sequence of feasible points with decreasing objective values. Here the algorithm to solve the problem Eqs. (3.57a) - (3.57g) is called sequential convex programming (SCP) with DC constraints [TDQ11], is applied.

For the problem Eq. (3.57a)-Eq. (3.57g), by leaving the convex objective function and all the convex constraints unchanged, and by linearizing $\boldsymbol{i}_{sz}^{*dq\top}\boldsymbol{Q}_{j1}\boldsymbol{i}_{sz}^{*dq}$ in Eq. (3.57e) and $\boldsymbol{i}_{sz}^{*dq\top}\boldsymbol{Q}_{CUF+}\boldsymbol{i}_{sz}^{*dq}$ in Eq. (3.57g) around the feasible point $\boldsymbol{i}_{sz}^{*dq(k)}$, the following optimal problem as Eq. (3.58a) - (3.58e) is derived,

$$min \quad f_{obj}(\boldsymbol{i}_{sz}^{*dq}) = \boldsymbol{i}_{sz}^{*dq\top}\boldsymbol{Q}_{obj}\boldsymbol{i}_{sz}^{*dq} + \boldsymbol{F}_{obj}\boldsymbol{i}_{sz}^{*dq} \tag{3.58a}$$

$$s.t. \quad \boldsymbol{i}_{sz}^{*dq} \in \Omega \tag{3.58b}$$

$$\boldsymbol{i}_{sz}^{*dq\top}\boldsymbol{Q}_{jH}\boldsymbol{i}_{sz}^{*dq} \leq 2\boldsymbol{i}_{sz}^{*dq(k)\top}\boldsymbol{Q}_{j1}(\boldsymbol{i}_{sz}^{*dq} - \boldsymbol{i}_{sz}^{*dq(k)}) + \boldsymbol{i}_{sz}^{*dq(k)\top}\boldsymbol{Q}_{j1}\boldsymbol{i}_{sz}^{*dq(k)} \tag{3.58c}$$

$$\boldsymbol{i}_{sz}^{*dq\top}\boldsymbol{Q}'_{jH}\boldsymbol{i}_{sz}^{*dq} \leq 2\boldsymbol{i}_{sz}^{*dq(k)\top}\boldsymbol{Q}'_{j1}(\boldsymbol{i}_{sz}^{*dq} - \boldsymbol{i}_{sz}^{*dq(k)}) + \boldsymbol{i}_{sz}^{*dq(k)\top}\boldsymbol{Q}'_{j1}\boldsymbol{i}_{sz}^{*dq(k)} \tag{3.58d}$$

$$\boldsymbol{i}_{sz}^{*dq\top}\boldsymbol{Q}_{CUF-}\boldsymbol{i}_{sz}^{*dq} \leq 2\boldsymbol{i}_{sz}^{*dq(k)\top}\boldsymbol{Q}_{CUF+}(\boldsymbol{i}_{sz}^{*dq} - \boldsymbol{i}_{sz}^{*dq(k)}) + \boldsymbol{i}_{sz}^{*dq(k)\top}\boldsymbol{Q}_{CUF+}\boldsymbol{i}_{sz}^{*dq(k)} \tag{3.58e}$$

in which, Ω is the non-empty closed convex set defined by the linear and convex quadratic constraints of Eq. (3.57b) - (3.57d). If we denote

$$\begin{aligned} D = \{\boldsymbol{i}_{sz}^{*dq} \in \Omega | \boldsymbol{i}_{sz}^{*dq\top}\boldsymbol{Q}_{jH}\boldsymbol{i}_{sz}^{*dq} - \boldsymbol{i}_{sz}^{*dq\top}\boldsymbol{Q}_{j1}\boldsymbol{i}_{sz}^{*dq} \leq 0, \\ \boldsymbol{i}_{sz}^{*dq\top}\boldsymbol{Q}'_{jH}\boldsymbol{i}_{sz}^{*dq} - \boldsymbol{i}_{sz}^{*dq\top}\boldsymbol{Q}'_{j1}\boldsymbol{i}_{sz}^{*dq} \leq 0, \end{aligned}$$

$$\boldsymbol{i}_{sz}^{*dq\top}\boldsymbol{Q}_{CUF-}\boldsymbol{i}_{sz}^{*dq} - \boldsymbol{i}_{sz}^{*dq\top}\boldsymbol{Q}_{CUF+}\boldsymbol{i}_{sz}^{*dq} \leq 0\} \tag{3.59}$$

as feasible set, the SCP algorithm can be described as following:

Initialization Set $k = 0$ and choose an initial feasible point $\boldsymbol{i}^{*dq(0)_{sz}} \in D$.
Iteration

Step 1 Solve the convex problem of Eq. (3.58a) - (3.58e) to obtain a solution $\boldsymbol{i}_{sz}^{*dq(k+1)}$. This step can be achieved by appropriate solvers.

Step 2 If $\left\|\boldsymbol{i}_{sz}^{*dq(k+1)} - \boldsymbol{i}_{sz}^{*dq(k)}\right\| \leq \boldsymbol{\epsilon}$ for a proper tolerance $\boldsymbol{\epsilon} \succ \mathbf{0}$ then terminate. Otherwise, set $k := k + 1$ and go to 1).

3.5. Sequential Convex Programming and Its Convergence

This section focuses on the convergence analysis of the following general linearised optimization problem,

$$\begin{cases} min & f_{obj}^{(k)}(\boldsymbol{x}) = f(\boldsymbol{x}) + \frac{\rho}{2}\left\|\boldsymbol{x} - \boldsymbol{x}^{(k)}\right\|_2^2 \\ s.t. & u_i(\boldsymbol{x}) - v_i(\boldsymbol{x}^{(k)}) - \nabla v_i(\boldsymbol{x}^{(k)})(\boldsymbol{x} - \boldsymbol{x}^{(k)}) \leq 0, \quad i = 1, 2, \cdots, \ell \\ & \boldsymbol{x} \in \Omega \end{cases} \tag{3.60}$$

where $\boldsymbol{x}$ is the optimization variable to be determined, $f(\boldsymbol{x})$ the original convex objective function, u_i and v_i, $i = (1, 2, \cdots, \ell)$ are convex functions. The regulation term $\frac{\rho}{2}\left\|\boldsymbol{x} - \boldsymbol{x}^{(k)}\right\|_2^2$ is added for strict convergence of the strict function with $\rho \geq 0$, however, it is not always needed and can be chosen small depending on the gradient of the objective function [DVLP+13]. In the optimization problem Eq. (3.58e), there exist $\rho = 0$.

Suppose that λ_i is the KKT multiplier. The KKT condition of the convex problem

Eq. (3.60) is as following,

$$\begin{cases} u_i(\boldsymbol{x}^{(k+1)}) - v_i(\boldsymbol{x}^{(k)}) - \nabla v_i(\boldsymbol{x}^{(k)})(\boldsymbol{x}^{(k+1)} - \boldsymbol{x}^{(k)}) \leq 0 \\ \lambda_i^{(k+1)} \geq 0 \\ \lambda_i^{(k+1)}[u_i(\boldsymbol{x}^{(k+1)}) - v_i(\boldsymbol{x}^{(k)}) - \nabla v_i(\boldsymbol{x}^{(k)})(\boldsymbol{x}^{(k+1)} - \boldsymbol{x}^{(k)})] = 0 \\ -\left(\nabla f(\boldsymbol{x}^{(k+1)}) + \rho(\boldsymbol{x}^{(k+1)} - \boldsymbol{x}^{(k)})\right) = \\ \qquad \sum_{i=1}^{\ell} \lambda_i^{(k+1)}[\nabla u_i(\boldsymbol{x}^{(k+1)}) - \nabla v_i(\boldsymbol{x}^{(k)})] + N_\Omega(\boldsymbol{x}^{(k+1)}) \end{cases} \tag{3.61}$$

in which, N_Ω is the multivalued mapping of the normal cone of Ω.

Based on the KKT conditions, we can obtain the next lemma giving a key property to prove the convergence of algorithm.

Lemma Suppose that f, u_i, v_i are ρ_f, ρ_{u_i} and ρ_{v_i}-convex [TDQ11] ($\rho_f \geq 0$, $\rho_{u_i} \geq 0$ and $\rho_{v_i} \geq 0$), then the sequence $\left\{(x^{(k)}), \lambda^{(k)}\right\}$ generated by the algorithm satisfies

$$\begin{aligned} f(\boldsymbol{x}^{(k)}) - f(\boldsymbol{x}^{(k+1)}) \geq & \frac{1}{2}(\rho_f + \sum_{i=1}^{\ell} \rho_{u_i}\lambda_i^{(k+1)}) \left\|\boldsymbol{x}^{(k)} - \boldsymbol{x}^{(k+1)}\right\|_2^2 \\ & + \frac{1}{2}\sum_{i=1}^{\ell} \rho_{v_i}\lambda_i^{(k+1)} \left\|\boldsymbol{x}^{(k)} - \boldsymbol{x}^{(k-1)}\right\|_2^2 + \rho \left\|\boldsymbol{x}^{(k)} - \boldsymbol{x}^{(k+1)}\right\|_2^2 \end{aligned} \tag{3.62}$$

It can be easily obtained by the proves from [DGMD12, TDQ11].

The properties of the algorithm can be summarized as:

1. Suppose that f is bounded from below on D, then $\lim_{k\to\infty} \left\|\boldsymbol{x}^{(k)} - \boldsymbol{x}^{(k+1)}\right\| = 0$. If the set of KKT point is finite, the whole sequence $\left\{(\boldsymbol{x}^{(k)}), \lambda^{(k)}\right\}$ convergences to a KKT point.

2. For a continuous function, there exist infinite possibilities to decompose it to the difference of two convex functions. Different selection of the u_i and v_i results in different ρ_{u_i} and ρ_{v_i}, which affect the convergence rate.

3. Even if f, u_i, v_i are all convex, but not strong convex, which means that $\rho_f = 0$, $\rho_{u_i} = 0$ and $\rho_{v_i} = 0$, there exists $f(\boldsymbol{x}^{(k+1)}) - f(\boldsymbol{x}^{(k)}) \leq -\rho \left\|\boldsymbol{x}^{(k)} - \boldsymbol{x}^{(k+1)}\right\|_2^2$ $\prec 0$ to make the objective function converge.

3.6. Block Diagram for Strategy Validation

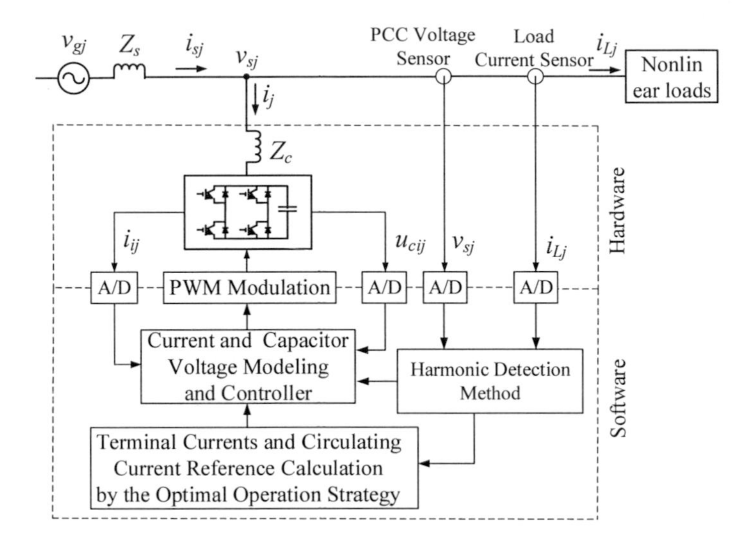

Figure 3.3.: Block diagram for the operation strategy validation: shunt connection of an active power filter to the line which feeds a nonlinear load. The active power filter applied here is delta-connected CHB, which is for simplicity represented by a H-bridge cell with a capacitor.

The subject of this chapter is the proposed operation strategy to determine the terminal currents and circulating current with apparent power reduction, without the blocks such as the modelling and controller design the strategy cannot be validated. Fig. 3.3 shows the application of a shunt APF based on delta-connected CHB without transformer to the power system which includes the hardware and software configurations, both of which are important for harmonic mitigation performance. At the PCC, the grid supply v_g that has associated impedance Z_s feeds the nonlinear load, which is connected in parallel to the APF. It assume that the harmonic detection works accurately, with the detected PCC voltages and load currents, the reference terminal currents and circulating current then can be determined via the proposed current strategy. At the same time, the calculated reference currents and the detected information are provided to modelling and controller block, whose output will be compared with the carriers to generate stepped multilevel waveforms.

A large number of different modulation algorithms have been adopted or developed depending on the application and the converter topology, each one having unique advantages and drawbacks. The most common modulation methods for multilevel converters are Phase-Shifted PWM (PSPWM), Level-Shifted PWM (LSPWM), and space vector modulation. Also other methods such as selective harmonic elimination modulation have been developed. The PSPWM modulation which has the advantage of even power distribution is used in this thesis. The switching of the SMs is based on PSPWM modulation which introduces N-multiple carrier frequency with associated sideband components [LYX+15]. The effect of this high frequency distortion can be, however, neglected as far as the carrier frequency is chosen high enough and will not cause performance degradation of the proposed optimal strategy.

The modelling and controller design will be introduced later, where the averaged model is used here and a periodic Linear Quadratic Regulator (LQR) controller is designed according to the derived model. It should be clear that the modelling and controller are introduced for validation, other modelling methods and controllers can be applied as well. Various modelling techniques have been reviewed in [RFK+09].

3.6.1. Harmonic Detection Method

The harmonic detection method has the task of detecting the harmonic currents/voltages that have to be used in other blocks. Errors given by the harmonic detection method degrade the overall performance of the APF. The literature has put forward many harmonic detection methods, which can be classified into time- and frequency-domain. Frequency-domain methods such as Fast Fourier Transform (FFT) and Discrete Fourier Transform (DFT), and time-domain methods such as synchronous fundamental dq frame as well as synchronous harmonic

dq frame are quite popular for harmonic detection [ABH07]. Discrete Fourier Transform is a mathematical transformation for discrete signals, giving both the amplitude and phase informations of the selected harmonic. The synchronous fundamental/harmonic dq frame method rotates with an angular speed of the selected fundamental/harmonic frequency so that only the respective frequency is dc component and others are ac components. The detection of the selected frequency is done with low-pass filters.

Other detection methods, for example, Wavelet filter and Kalman filter can also be found in literature. Among them, Kalman filter, belonging to time-domain methods, works recursively to produce precise estimation by using noisy incoming data. The author of this thesis also did some research on an adaptive Kalman filter methodology for dynamic harmonic detection to achieve a good accuracy and dynamic performance, see [WL15], assuming that the noises come from sensor inaccuracies.

This chapter uses the synchronous dq frame method for harmonic detection because the algorithm is simple and during the simulation phase there does not exist noises.

3.6.2. Modelling and Controller Design

Continuous Time Model

When a continuous time average model is used, the piece-wise feature of switches is approximated with an overall smooth trait and balancing between the submodule capacitor voltage is assumed, the submodules are replaced by an equivalent voltage source. It should be noticed that the continuous switching function s_{ij} is the low-frequency ($<2\,\mathrm{kHz}$) modulation signal, which will be used by each submodule in branch-ij to compare with the phase-shifted carrier signals to generate the stepped waveforms.

Define the branch currents and the sum of SM capacitor voltages in each branch (already defined in Eq. (3.12)) as state variables, the switching signals as control inputs and the PCC line-line voltages as disturbance for branch-ij, the continuous time model can be written as following when the power loss in the branch inductor and resistor is ignored,

$$\dot{\boldsymbol{x}} = \widetilde{\boldsymbol{A}}(t)\boldsymbol{x} + \widetilde{\boldsymbol{B}}\boldsymbol{u} + \widetilde{\boldsymbol{B}}_{dt}\boldsymbol{d} \tag{3.63}$$

where

$$\boldsymbol{x} = \begin{bmatrix} i_{ab} & i_{bc} & i_{ca} & u_{Cab} & u_{Cbc} & u_{Cca} \end{bmatrix}^\top \tag{3.64}$$

$$\boldsymbol{u} = \begin{bmatrix} s_{ab} & s_{bc} & s_{ca} \end{bmatrix}^\top \tag{3.65}$$

$$\boldsymbol{d} = \begin{bmatrix} v_{sab} & v_{sbc} & v_{sca} \end{bmatrix}^\top \tag{3.66}$$

$$\widetilde{\boldsymbol{A}}(t) = \begin{bmatrix} -\frac{r}{L} & 0 & 0 & 0 & 0 & 0 \\ 0 & -\frac{r}{L} & 0 & 0 & 0 & 0 \\ 0 & 0 & -\frac{r}{L} & 0 & 0 & 0 \\ \frac{1}{C_{sum}} s^*_{ab} & 0 & 0 & 0 & 0 & 0 \\ 0 & \frac{1}{C_{sum}} s^*_{bc} & 0 & 0 & 0 & 0 \\ 0 & 0 & \frac{1}{C_{sum}} s^*_{ca} & 0 & 0 & 0 \end{bmatrix} \tag{3.67}$$

$$\widetilde{\boldsymbol{B}} = \begin{bmatrix} -\frac{1}{L} u^*_{C,dc} & 0 & 0 \\ 0 & -\frac{1}{L} u^*_{C,dc} & 0 \\ 0 & 0 & -\frac{1}{L} u^*_{C,dc} \\ 0 & 0 & 0 \\ 0 & 0 & 0 \\ 0 & 0 & 0 \end{bmatrix} \tag{3.68}$$

$$\widetilde{\boldsymbol{B}}_{dt} = \begin{bmatrix} \frac{1}{L} & 0 & 0 \\ 0 & \frac{1}{L} & 0 \\ 0 & 0 & \frac{1}{L} \\ 0 & 0 & 0 \\ 0 & 0 & 0 \\ 0 & 0 & 0 \end{bmatrix} \tag{3.69}$$

It should be mentioned that the PSPWM modulation method combined with the sorting algorithm achieves the intrinsic balancing of submodule capacitor voltages by sorting those submodules which have high capacitor voltage in discharging mode and those with low capacitor voltage for the charging mode, therefore it does not require a voltage control loop for each individual submodule [RMB+15]. That's why Eq. (3.63) does not model the individual submodule capacitor voltage.

The desired branch current i^*_{ij} can be obtained from Eq. (3.27) when the desired terminal currents and circulating current achived from the proposed current strategy. The sum of SM capacitor voltages in each branch at steady state can be estimated as

$$u^*_{Cij} = u^*_{C,dc} + \frac{1}{C_{sum}u^*_{C,dc}} \int p^*_{ij} dt = u^*_{C,dc} + \frac{1}{C_{sum}u^*_{C,dc}} \int v_{sij} i^*_{ij} dt \tag{3.70}$$

where $u^*_{C,dc}$ is the dc component of u^*_{Cij}. The desired switching function in branch-ij is often estimated as the ratio between the desired branch voltage and the dc component of submodule capacitor sum voltages in the corresponding branch by neglecting capacitor voltage ripple [LWY+17, KV17], see follows

$$\boldsymbol{s}^*_{ij} = u^*_{ij}/u^*_{C,dc} \tag{3.71}$$

Discrete Time Model

Given the system presented in (3.63) and the sample period T_s, the system can be discretized to the equation in standard form [Gen06] as Eq. (3.72),

$$\boldsymbol{x}_{k+1} = \widetilde{\boldsymbol{\Phi}}_k \boldsymbol{x}_k + \widetilde{\boldsymbol{\Gamma}}_k \boldsymbol{u}_k + \widetilde{\boldsymbol{\Gamma}}_{dtk} \boldsymbol{d}_k \tag{3.72}$$

where $\widetilde{\boldsymbol{\Phi}}_k = T_s \widetilde{\boldsymbol{A}}(t_k) + \boldsymbol{I}$, $\widetilde{\boldsymbol{\Gamma}}_k = T_s \widetilde{\boldsymbol{B}}$ and $\widetilde{\boldsymbol{\Gamma}}_{dtk} = T_s \widetilde{\boldsymbol{B}}_{dt}$.

Controller Design in Discrete Time

It can be seen from Eq. (3.68) that $\widetilde{\boldsymbol{A}}(t)$ is a periodic time-varying matrix due to the time-varying s^*_{ij}. Eq. (3.63) is a periodic time-varying system. It is difficult to transform the time-varying system Eq. (3.63) to a time-invariant system such as in dq frame due to the product of time-varying switching functions and branch currents, both of which contain multiple line-frequency components. As a result, this chapter needs to tackle the control problem associated with the periodic time-varying system, which is quite challenging because the time-varying eigenvalues

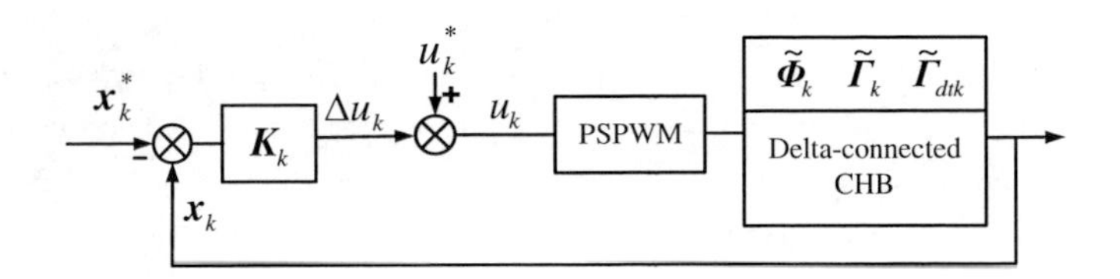

Figure 3.4.: Feedback control loop for the delta-connected CHB system

of the periodic matrix can not determine the stability of the system. There is an excellent survey of periodic systems and control in [SB99]. In this chapter the optimal periodic LQR controller is designed.

With the desired state variables calculated from Eq. (3.27) and Eq. (3.70), the desired control input from Eq. (3.71), and the definition $\Delta \boldsymbol{x}_k = \boldsymbol{x}_k - \boldsymbol{x}_k^*$ as well as $\Delta \boldsymbol{u}_k = \boldsymbol{u}_k - \boldsymbol{u}_k^*$, a periodic optimal feedback controller (finite horizon discrete time LQR) [BBM17] is designed as follows

$$\Delta \boldsymbol{u}_k = -\boldsymbol{K}_k \Delta \boldsymbol{x}_k, \quad (k = 0, \ldots, k_f) \tag{3.73}$$

where $\boldsymbol{K}_0 = \boldsymbol{K}_{k_f}$, to minimize the performance index

$$min \quad J(\Delta \boldsymbol{U}) = \sum_{k=0}^{k_f - 1} \left(\Delta \boldsymbol{x}_k^T \boldsymbol{Q} \Delta \boldsymbol{x}_k + \Delta \boldsymbol{u}_k^T \boldsymbol{R} \Delta \boldsymbol{u}_k \right) + \Delta \boldsymbol{x}_{k_f}^T \boldsymbol{Q} \Delta \boldsymbol{x}_{k_f} \tag{3.74a}$$

$$s.t. \qquad \Delta \boldsymbol{x}_{k+1} = \widetilde{\boldsymbol{\Phi}}_k \Delta \boldsymbol{x}_k + \widetilde{\boldsymbol{\Gamma}}_k \Delta \boldsymbol{u}_k \tag{3.74b}$$

where $\Delta \boldsymbol{U} = (\Delta \boldsymbol{u}_0, \ldots, \Delta \boldsymbol{u}_{k_f - 1})$, both the error weighting matrix $\boldsymbol{Q}$ and the control weighting matrix $\boldsymbol{R}$ are symmetric and positive definite. The detailed controller design procedure will be introduced in Section 3.6.3. The feedback system under consideration with the control loop is shown in Fig. 3.4.

3.6.3. Finite Horizon Discrete Time LQR via Dynamic Programming

It will present the approach to solve the problem Eq. (3.74a)-(3.74b). Two alternate approaches have been introduced in [BBM17]. One is batch approach, which is solved as a least-square problem, leading to big matrix inversion calaculation, since the batch approach requires the repeated inversion of a potentially large matrix. Another is recursive approach using dynamic programming. The dynamic programming approach is a more efficient way to generate the feedback policy, i.e., a sequence of feedback laws expressing at each time step, the control action as a function of the state at that time. In the dynamic programming method, it tackles the optimization problem Eq. (3.77a) - (3.77b).

Steps of Dynamic Programming

1. Set $\boldsymbol{P}_{k_f} = \boldsymbol{Q}$. This reflects the knowledge that on the final time step, with no time left the cost-to-go is just final state cost.

2. For $k = k_f, k_f - 1, \ldots, 1$

$$\begin{aligned}\boldsymbol{P}_{k-1} = {} & \widetilde{\boldsymbol{\Phi}}_{k-1}^T \boldsymbol{P}_k \widetilde{\boldsymbol{\Phi}}_{k-1} + \boldsymbol{Q} \\ & - (\widetilde{\boldsymbol{\Phi}}_{k-1}^T \boldsymbol{P}_k \widetilde{\boldsymbol{\Gamma}}_{k-1})(\boldsymbol{R} + \widetilde{\boldsymbol{\Gamma}}_{k-1}^T \boldsymbol{P}_k \widetilde{\boldsymbol{\Gamma}}_{k-1})^{-1}(\widetilde{\boldsymbol{\Gamma}}_{k-1}^T \boldsymbol{P}_k \widetilde{\boldsymbol{\Phi}}_{k-1}) \end{aligned} \quad (3.75)$$

   This is known as the Differential Riccati Equation.

3. For $k = 0, 1, \ldots, k_f - 1$, the control gain is calculated as

$$\boldsymbol{K}_k = (\boldsymbol{R} + \widetilde{\boldsymbol{\Gamma}}_{k-1}^T \boldsymbol{P}_k \widetilde{\boldsymbol{\Gamma}}_{k-1})^{-1}(\widetilde{\boldsymbol{\Gamma}}_{k-1}^T \boldsymbol{P}_k \widetilde{\boldsymbol{\Phi}}_{k-1}) \quad (3.76)$$

4. For $k = 0, 1, \ldots, k_f - 1$, the optimal control input is $\Delta \boldsymbol{u}_k = -\boldsymbol{K}_k \Delta \boldsymbol{x}_k$.

Provement

According to Bellman's principle of optimality, no matter what the initial state and initial decision are, the remaining decisions must constitute an optimal policy with regard to the state resulting from the first decision [WHK05].

For $k = 0, \ldots, k_f$, define k-horizon value function

$$V_k(\Delta \boldsymbol{x}_k) = min \left\{ \sum_{i=k}^{k_f-1} (\Delta \boldsymbol{x}_i^T \boldsymbol{Q} \Delta \boldsymbol{x}_i + \Delta \boldsymbol{u}_i^T \boldsymbol{R} \Delta \boldsymbol{u}_i) + \Delta \boldsymbol{x}_{k_f}^T \boldsymbol{Q} \Delta \boldsymbol{x}_{k_f} \right\} \quad (3.77a)$$

$$s.t. \qquad \Delta \boldsymbol{x}_{i+1} = \widetilde{\boldsymbol{\Phi}}_i \Delta \boldsymbol{x}_i + \widetilde{\boldsymbol{\Gamma}}_i \Delta \boldsymbol{u}_i, \quad (i = k, \ldots, k_f) \tag{3.77b}$$

where $V_k(\Delta \boldsymbol{x}_k)$ gives the minimum LQR cost-to-go, starting from state $\Delta \boldsymbol{x}_k$ at time k.

When no time left is only final state cost, $V_{k_f} = \Delta \boldsymbol{x}_{k_f}^T \boldsymbol{Q} \Delta \boldsymbol{x}_{k_f}$ and

$$\begin{aligned}
&V_{k-1}(\Delta \boldsymbol{x}_{k-1}) \\
&= min \left\{ \Delta \boldsymbol{x}_{k-1}^T \boldsymbol{Q} \Delta \boldsymbol{x}_{k-1} + \Delta \boldsymbol{u}_{k-1}^T \boldsymbol{R} \Delta \boldsymbol{u}_{k-1} + V_k(\Delta \boldsymbol{x}_k) \right\} \\
&= min \left\{ \Delta \boldsymbol{x}_{k-1}^T \boldsymbol{Q} \Delta \boldsymbol{x}_{k-1} + \Delta \boldsymbol{u}_{k-1}^T \boldsymbol{R} \Delta \boldsymbol{u}_{k-1} \right. \\
&\qquad \left. + (\widetilde{\boldsymbol{\Phi}}_{k-1} \Delta \boldsymbol{x}_{k-1} + \widetilde{\boldsymbol{\Gamma}}_{k-1} \Delta \boldsymbol{u}_{k-1})^T P_k (\widetilde{\boldsymbol{\Phi}}_{k-1} \Delta \boldsymbol{x}_{k-1} + \widetilde{\boldsymbol{\Gamma}}_{k-1} \Delta \boldsymbol{u}_{k-1}) \right\} \\
&= min \begin{bmatrix} \Delta \boldsymbol{x}_{k-1} \\ \Delta \boldsymbol{u}_{k-1} \end{bmatrix}^T \begin{bmatrix} \boldsymbol{Q} + \widetilde{\boldsymbol{\Phi}}_{k-1}^T P_k \widetilde{\boldsymbol{\Phi}}_{k-1} & \widetilde{\boldsymbol{\Gamma}}_{k-1}^T P_k \widetilde{\boldsymbol{\Phi}}_{k-1} \\ \widetilde{\boldsymbol{\Phi}}_{k-1}^T P_k \widetilde{\boldsymbol{\Gamma}}_{k-1} & \boldsymbol{R} + \widetilde{\boldsymbol{\Gamma}}_{k-1}^T P_k \widetilde{\boldsymbol{\Gamma}}_{k-1} \end{bmatrix} \begin{bmatrix} \Delta \boldsymbol{x}_{k-1} \\ \Delta \boldsymbol{u}_{k-1} \end{bmatrix}
\end{aligned} \tag{3.78}$$

Eq. (3.78) can be written as

$$\begin{aligned}
&V_{k-1}(\Delta \boldsymbol{x}_{k-1}) \\
&= min \left\{ \Delta \boldsymbol{x}_{k-1}^T (\boldsymbol{Q} + \widetilde{\boldsymbol{\Phi}}_{k-1}^T P_k \widetilde{\boldsymbol{\Phi}}_{k-1} - \boldsymbol{K}_{k-1}^T (\boldsymbol{R} + \widetilde{\boldsymbol{\Gamma}}_{k-1}^T P_k \widetilde{\boldsymbol{\Gamma}}_{k-1}) \boldsymbol{K}_{k-1}) \Delta \boldsymbol{x}_{k-1} \right. \\
&\quad \left. + (\Delta \boldsymbol{u}_{k-1} + \boldsymbol{K}_{k-1} \Delta \boldsymbol{x}_{k-1})^T (\boldsymbol{R} + \widetilde{\boldsymbol{\Gamma}}_{k-1}^T P_k \widetilde{\boldsymbol{\Gamma}}_{k-1}) (\Delta \boldsymbol{u}_{k-1} + \boldsymbol{K}_{k-1} \Delta \boldsymbol{x}_{k-1}) \right\}
\end{aligned} \tag{3.79}$$

where $\boldsymbol{K}_{k-1}$ is defined in Eq. (3.76). When $\Delta \boldsymbol{u}_{k-1} = -\boldsymbol{K}_{k-1} \Delta \boldsymbol{x}_{k-1}$, Eq. (3.79) is minimized to the following expression,

$$V_{k-1}(\Delta \boldsymbol{x}_{k-1}) = \Delta \boldsymbol{x}_{k-1}^T \boldsymbol{P}_{k-1} \Delta \boldsymbol{x}_{k-1} \tag{3.80}$$

where $\boldsymbol{P}_{k-1}$ is as Eq. (3.75).

3.7. Validation via Time-Domain Simulation

Table 3.1 lists the parameters used in the optimization problem Eq. (3.58a)-Eq. (3.58e). In order to solve the convex quadratically constrained quadratic programming subproblems in **step 1** of the SCP algorithm, the IBM ILOG CPLEX Optimizer is chosen, which provides the interface with MATLAB.

Table 3.1.: Parameters used in optimization

total harmonic distortion upper bound	λ_j	5%
individual harmonic distortion upper bound	λ_{jH}	4%
fundamental current unbalance upper bound	γ_1	5%
power factor lower bound	$\underline{pf}$	0.8
tolerance	$\boldsymbol{\epsilon}$	$0.1\boldsymbol{I}_{6M\times 1}$

Different simulation studies have been conducted in MATLAB/Simulink version 2012a to evaluate the proposed strategy. The simulation parameters are summarized in Table 3.2. The dc component of the sum capacitor voltages of each branch is set to $u^*_{C,dc}$ =62 kV, the initial value of u_{Cij} to 40 kV.

Table 3.2.: Parameters used in simulations

Rated line frequency	$\omega/2\pi$	50 Hz
DC capacitor of full-bridge submodule	C	240 μF
Carrier frequency for PSPWM	f_c	5000 Hz
Branch inductor	L	9.2 mH
Branch resistor	r	0.0566 Ω
Number of SMs in one branch	N	8
Sampling time	T_s	0.000005
Error weighted matrix	$\boldsymbol{Q}$	$\begin{bmatrix} 0.001\boldsymbol{I}_3 & 0 \\ 0 & 0.0000002\boldsymbol{I}_3 \end{bmatrix}$
Control weighted matrix	$\boldsymbol{R}$	$\boldsymbol{I}_3$

Four cases are presented here,

Case A ideal PCC voltages

Case B unbalanced and undistorted PCC voltages

Case C balanced and distorted PCC voltages

Case D unbalanced and distorted PCC voltages

In all cases the load currents in KA are as

$$\begin{cases} i_{La} = \sin\left(\omega t - \frac{\pi}{18}\right) + 0.2\sin\left(5\omega t - \frac{\pi}{18}\right) + 0.14\sin\left(7\omega t - \frac{\pi}{18}\right) \\ i_{Lb} = \sin\left(\omega t - \frac{13\pi}{18}\right) + 0.2\sin\left(5\omega t + \frac{11\pi}{18}\right) + 0.14\sin\left(7\omega t - \frac{13\pi}{18}\right) \\ i_{Lc} = \sin\left(\omega t + \frac{11\pi}{18}\right) + 0.2\sin\left(5\omega t - \frac{13\pi}{18}\right) + 0.14\sin\left(7\omega t + \frac{11\pi}{18}\right) \end{cases} \tag{3.81}$$

Table 3.3- 3.6 summarize various current strategies for Case A-D respectively. In these cases the following strategies are compared, Perfect Harmonic Cancellation (PHC), Unity Power Factor (UPF), the proposed optimal (OPT) strategy, current strategies (CST1-CST4) which satisfy constraints Eqs. (3.57b)-(3.57g) but with some of the operation condition slightly different from the calculated OPT strategy.

PHC: The PHC source currents can be expressed by

$$\begin{bmatrix} i_{sa}^* \\ i_{sb}^* \\ i_{sc}^* \end{bmatrix}_{PHC} = g_{PHC} \begin{bmatrix} v_{sa+1} \\ v_{sb+1} \\ v_{sc+1} \end{bmatrix}, \quad g_{PHC} = \frac{p_{L,dc}}{3V_{s+1}^2} \tag{3.82}$$

where V_{s+1} is the effective positive-sequence fundamental frequency component in PCC voltages, seen in Eq. (2.1), meanwhile there is $V_{s+1} = \sqrt{\left(v_{sd+1}^2 + v_{sq+1}^2\right)/3}$. g_{PHC} can be called PHC conductance. In the PHC strategy, the average active power transferred from the power supply is

$$p_{s,dc} = \begin{bmatrix} v_{sd+1} & v_{sq+1} \end{bmatrix} \begin{bmatrix} i_{sd+1}^* \\ i_{sq+1}^* \end{bmatrix}_{PHC} = g_{PHC}\left(v_{sd+1}^2 + v_{sq+1}^2\right) = p_{L,dc} \tag{3.83}$$

UPF: The UPF source currents can be expressed by

$$\begin{bmatrix} i_{sa}^* \\ i_{sb}^* \\ i_{sc}^* \end{bmatrix}_{UPF} = g_{UPF} \begin{bmatrix} \sum_n (v_{sa+n} + v_{sa-n}) \\ \sum_n (v_{sb+n} + v_{sb-n}) \\ \sum_n (v_{sc+n} + v_{sc-n}) \end{bmatrix}, \quad g_{UPF} = \frac{p_{L,dc}}{3\sum_n (V_{s+n}^2 + V_{s-n}^2)} \tag{3.84}$$

where V_{s+n} and V_{s-n} are the nth effective positive- and gegative-sequence component in PCC voltages, seen in Eq. (2.1). g_{UPF} can be called UPF conductance. In the UPF strategy, the average active power transferred from the power supply is

$$\begin{aligned} p_{s,dc} &= \sum_n \begin{bmatrix} v_{sd+n} & v_{sq+n} & v_{sd-n} & v_{sq-n} \end{bmatrix} \begin{bmatrix} i^*_{sd+n} \\ i^*_{sq+n} \\ i^*_{sd-n} \\ i^*_{sq-n} \end{bmatrix}_{UPF} \\ &= g_{UPF} \sum_n \left(v^2_{sd+n} + v^2_{sq+n} + v^2_{sd-n} + v^2_{sq-n} \right) = p_{L,dc} \end{aligned} \tag{3.85}$$

CST1: The total source current distortion, the frequency order of circulating current, and the source current unbalance factor in CST1 is same as OPT strategy, however, the individual source current distortion is different with OPT. CST1 is for Case A-D.

CST2-CST3: The individual and total source current distortion, the frequency order of circulating current in CST2 and CST3 strategies are the same as those in OPT, but the source current unbalance factor in CST2 and CST3 is respectively smaller and larger than OPT. CST2-CST3 are proposed only for Case B and D since there is unbalance only in these two cases.

CST4: The individual and total source current distortion, the source current unbalance factor in CST4 are the same as those in OPT, but the frequency order of circulating current in CST4 is different with OPT. CST4 is only for Case D since in this case there can be different circulating current references.

Table 3.3- 3.6 give the parameters, including the 5th and 7th individual source current distortion (IHD) of each phase, source current total harmonic distortion (THD_j, $j = a, b, c$), source current negative-sequence unbalance factor of the fundamental frequency (CUF), the frequency orders contained in the circulating current, power factor (pf), RMS of desired branch currents (I_e), average active power of the load ($p_{L,dc}$), and the APF apparent power S_{APF}.

According to the instantaneous power theory [HAA07], the active and reactive power definition can be presented in term of *abc*-frame,

$$p^*_{APF} = \underbrace{v_{sab} i^*_{ab}}_{p^*_{ab}} + \underbrace{v_{sbc} i^*_{bc}}_{p^*_{bc}} + \underbrace{v_{sca} i^*_{ca}}_{p^*_{ca}} \tag{3.86}$$

$$q^*_{APF} = \underbrace{\frac{1}{\sqrt{3}}(v_{sbc} - v_{sca})i^*_{ab}}_{q^*_{ab}} + \underbrace{\frac{1}{\sqrt{3}}(v_{sca} - v_{sab})i^*_{bc}}_{q^*_{bc}} + \underbrace{\frac{1}{\sqrt{3}}(v_{sab} - v_{sbc})i^*_{ca}}_{q^*_{ca}} \quad (3.87)$$

These two powers have constant values and a superposition of oscillating power components.

$$\begin{aligned} p^*_{APF} &= \underset{\text{Average powers}}{p^*_{APF,dc}} + \underset{\text{Oscillating powers}}{\widetilde{p}^*_{APF}} && (3.88) \\ q^*_{APF} &= q^*_{APF,dc} + \widetilde{q}^*_{APF} && (3.89) \end{aligned}$$

Average powers — Oscillating powers

where $p^*_{APF,dc}$ and $\widetilde{p}^*_{APF}$ represent the average and oscillating part of p^*_{APF}, whereas $q^*_{APF,dc}$ and $\widetilde{q}^*_{APF}$ represent the average and oscillating part of q^*_{APF}. The oscillating powers contain multiple line-frequency components due to the multiplication of distorted and unbalanced PCC line-line voltages and branch currents. Moreover,

$$p^*_{APF,dc} = \frac{1}{T}\int_0^T p^*_{APF}dt, \qquad q^*_{APF,dc} = \frac{1}{T}\int_0^T q^*_{APF}dt \quad (3.90)$$

$$\widetilde{p}^*_{APF,e} = \sqrt{\frac{1}{T}\int_0^T \left[\widetilde{p}^*_{APF}\right]^2 dt}, \quad \widetilde{q}^*_{APF,e} = \sqrt{\frac{1}{T}\int_0^T \left[\widetilde{q}^*_{APF}\right]^2 dt} \quad (3.91)$$

It should be noticed that the APF apparent power S_{APF} satisfies

$$S_{APF} = 3V_{slle}I_e \quad (3.92)$$

$$S^2_{APF} = \underbrace{p^{*2}_{APF,dc} + \widetilde{p}^{*2}_{APF,e}}_{P^2} + \underbrace{q^{*2}_{APF,dc} + \widetilde{q}^{*2}_{APF,e}}_{Q^2} + D^2 \quad (3.93)$$

where V_{slle} is the effective PCC line-line voltage, I_e is as Eq. (3.32), D is called distortion power. Eq. (3.93) is used for the analysis of circuit systems under nonsinusoidal conditions.

In order to give insight into the detailed power components generated by each branch in the delta-connected CHB based shunt APF, Tables 3.7 - 3.8 give various APF power components of under PHC, UPF, CST1 and OPT strategies, including

- the nth order effective oscillating active and reactive power, denoted by P_{en} and Q_{en}, $P_{en} = \sqrt{P_{abn,e} + P_{bcn,e} + P_{can,e}/3}$, $P_{ijn,e} = \sqrt{\frac{1}{T}\int_0^T \left[p^*_{ijn}\right]^2 dt}$;

$Q_{en} = \sqrt{Q_{abn,e} + Q_{bcn,e} + Q_{can,e}/3}$, $Q_{ijn,e} = \sqrt{\frac{1}{T}\int_0^T \left[q^*_{ijn}\right]^2 dt}$. The powers are impacted by the interaction of PCC line-line voltages and branch currents. This interaction analysis will be introduced in the next chapter. How P_{en} and Q_{en} can be calculated via the interaction analysis technique can be seen in Section A.1.

- the effective value of oscillating active power and reactive power, denoted by P_e and Q_e, which are calculated from $P_e = \sqrt{\sum_n P_{en}^2}$ and $Q_e = \sqrt{\sum_n Q_{en}^2}$.

3.7.1. Case A

The ideal PCC voltages (in kV) are assumed to be Eq. (3.94). The load currents are as Eq. (3.81) (KA).

$$\begin{cases} v_{sa} = & 30.4\sin(\omega t) \\ v_{sb} = & 30.4\sin\left(\omega t - \frac{2\pi}{3}\right) \\ v_{sc} = & 30.4\sin\left(\omega t + \frac{2\pi}{3}\right) \end{cases} \tag{3.94}$$

Perfect Harmonic Cancellation (PHC) and Unity Power Power (UPF) can be achieved simultaneously, when PCC voltages are ideal. Under the OPT strategy, in order to minimize the APF apparent power, the source currents contain 5th and 7th order harmonics, the 5th and 7th harmonic distortion are respectively 4.0% and 3.0%, and the power factor is less than unity while meeting the power quality constraints, see Table 3.3. Under the CST1 strategy, the 5th and 7th harmonic distortion in source currents are 3.0% and 4.0% respectively, which are different with those in the OPT strategy, though the CST1 strategy has the same total harmonic distortion THD_j and unbalance factor CUF in source currents as those of our OPT strategy, its APF apparent power is larger than that of the OPT strategy.

It can also be seen from Table 3.7 that the APF average active power $p^*_{APF,dc} = 0$ and average reactive power $q^*_{APF,dc}$ is same for all the strategies. The effective oscillating active and reactive power P_e and Q_e in the OPT strategy are smaller than other strategies, validating that the power consumption of APF is minimum under the OPT strategy.

Fig. 3.6 shows the simulation results of the control of APF with reference currents calculated from the OPT strategy. The shunt APF is inserted at $t = 0.04s$, it can be seen from Fig. 3.6 that before compensation the source currents are severely distorted, after compensation, the source currents are slightly distorted, and the circulating current at steady state is zero in this case; the capacitor voltages are regulated to 62 kV. The desired current tracking and capacitor voltage regulation are achieved thanks to the effective controller. Fig. 3.10 (a) shows the RMS amplitude spectrum of the source currents before and after compensation. Besides the RMS values of three-phase source currents in each frequency order, it can be observed from this figure if there exists source current unbalance. When the nth order source currents are balanced, $I_{san} = I_{sbn} = I_{scn}$ should be satisfied.

Table 3.3.: Result summary of Case A

case		A			
strategy		PHC	UPF	CST1	**OPT**
5th IHD (%)	a	0	0	3.0	**4.0**
	b	0	0	3.0	**4.0**
	c	0	0	3.0	**4.0**
7th IHD (%)	a	0	0	4.0	**3.0**
	b	0	0	4.0	**3.0**
	c	0	0	4.0	**3.0**
THD_i (%)	a	0	0	5.0	**5.0**
	b	0	0	5.0	**5.0**
	c	0	0	5.0	**5.0**
CUF		0	0	0	**0**
Circulating Current		/	/	/	/
pf		1	1	0.9988	**0.9988**
I_e (A)		122.307	122.307	107.495	**106.575**
$p_{L,dc}$ (MW)		44.907	44.907	44.907	**44.907**
S_{APF} (MVA)		13.661	13.661	12.007	**11.904**

3.7.2. Case B

Here we consider a single-phase voltage sag as Eq.(3.95) in (kV), which has the zero- and negative-sequence voltage unbalance factor $V_{s01}/V_{s+1} = 33.3\%$ and $V_{s-1}/V_{s+1} = 33.3\%$. The load currents are as Eq. (3.81) (KA).

$$\begin{cases} v_{sa} = & 7.6\sin(\omega t) \\ v_{sb} = & 30.4\sin\left(\omega t - \frac{2\pi}{3}\right) \\ v_{sc} = & 30.4\sin\left(\omega t + \frac{2\pi}{3}\right) \end{cases} \tag{3.95}$$

The desired source currents under the PHC strategy are free of harmonics and balanced. In the UPF strategy the current unbalance factor is also 33.3% and the

Table 3.4.: Result summary of Case B

case		B					
strategy		PHC	UPF	CST1	CST2	CST3	**OPT**
5th IHD (%)	a	0	0	3.0	4.0	4.0	**4.0**
	b	0	0	3.0	4.0	4.0	**4.0**
	c	0	0	3.0	4.0	4.0	**4.0**
7th IHD (%)	a	0	0	4.0	3.0	3.0	**3.0**
	b	0	0	4.0	3.0	3.0	**3.0**
	c	0	0	4.0	3.0	3.0	**3.0**
THD_j (%)	a	0	0	5.0	5.0	5.0	**5.0**
	b	0	0	5.0	5.0	5.0	**5.0**
	c	0	0	5.0	5.0	5.0	**5.0**
CUF (%)		0	33.3	0.2	0	5.0	**0.2**
Circulating Current		1st	1st	1st	1st	1st	**1st**
pf		0.9045	0.9535	0.9040	0.9034	0.9173	**0.9028**
I_e (A)		127.339	197.092	113.202	112.315	115.223	**112.310**
$p_{L,dc}$ (MW)		33.680	33.680	33.680	33.680	33.680	**33.680**
S_{APF} (MVA)		11.245	17.404	9.996	9.918	10.174	**9.917**

power factor in UPF is less than 1 due to that there is zero-sequence component in the PCC voltages but not in the UPF source currents. Due to the unbalanced PCC voltages, the desired circulating current is not 0 but of fundamental frequency. It can be seen in Table 3.4, the desired source currents under the CST1 strategy contain the same THD_j and unbalance factor as the OPT strategy, but the individual harmonic distortion of CST1 and OPT are different, the results show that the APF apparent power of CST1 is larger than OPT. CST2 and CST3 have the same individual harmonic distortion as OPT, seen in Table 3.4, but with a smaller and larger source current unbalance factor (CUF) respectively than OPT strategies: in the CST2 strategy the CUF is 0, in the CST3 strategy the CUF is 5%, and in the OPT strategy 0.2%, both CST2 and CST3 have a larger APF apparent power than OPT. It should be noticed that though the apparent power of CST2 is very slightly greater than that of OPT due to that they have very close operation current condition, it illustrates that the proposed algorithm can lead to an optimal solution in mathematical respect.

Table 3.7 gives the power components of the PHC, UPF, CST1 and OPT strategy, showing that P_e and Q_e of the OPT strategy, are smaller than other strategies, showing APF power consumption of OPT is minimum. Fig. 3.7 gives the simulation results of the APF control under the OPT strategy, showing that the reference currents are tracked and the SM capacitor voltages are regulated to the reference level. The RMS amplitude spectrum of source currents before and after compensation by various strategies has been shown in Fig. 3.10 (b), from which the fundamental frequency source current unbalance can be obviously observed in UPF and CST3, since I_{sa1} obviously smaller than I_{sb1} and I_{sc1} in these two strategies. The unbalance in OPT cannot obviously seen from Fig. 3.10 (b) due to a very small CUF in the OPT strategy.

3.7.3. Case C

The PCC voltages as Eq. (3.96) in kV are balanced and contain the 5th and 7th harmonics. The load currents are as Eq. (3.81) (KA).

$$\begin{cases} v_{sa} = 30.4\sin(\omega t) + 5\sin(5\omega t) + 2\sin(7\omega t) \\ v_{sb} = 30.4\sin\left(\omega t - \frac{2\pi}{3}\right) + 5\sin\left(5\omega t + \frac{2\pi}{3}\right) + 2\sin\left(7\omega t - \frac{2\pi}{3}\right) \\ v_{sc} = 30.4\sin\left(\omega t + \frac{2\pi}{3}\right) + 5\sin\left(5\omega t - \frac{2\pi}{3}\right) + 2\sin\left(7\omega t + \frac{2\pi}{3}\right) \end{cases} \tag{3.96}$$

The PHC source currents are in phase with the fundamental frequency of the PCC voltages. The UPF source currents have the same 5th, 7th individual harmonic

Table 3.5.: Result summary of Case C

case		C			
strategy		PHC	UPF	CST1	**OPT**
5th IHD (%)	a	0	16.45	3.0	**4.0**
	b	0	16.45	3.0	**4.0**
	c	0	16.45	3.0	**4.0**
7th IHD (%)	a	0	6.58	4.0	**3.0**
	b	0	6.58	4.0	**3.0**
	c	0	6.58	4.0	**3.0**
THD_j (%)	a	0	17.71	5.0	**5.0**
	b	0	17.71	5.0	**5.0**
	c	0	17.71	5.0	**5.0**
CUF		0	0	0	**0**
Circulating Current		/	/	/	/
pf		0.9847	1	0.9909	**0.9919**
I_e (A)		123.473	80.008	108.179	**107.211**
$p_{L,dc}$ (MW)		46.798	46.798	46.798	**46.798**
S_{APF} (MVA)		14.006	9.076	12.271	**12.162**

distortion, and total harmonic distortion as the PCC voltages, which are 16.45%, 6.58%, and 17.71%, violating the current harmonic limits as per IEEE STD-519, although UPF leads to a smaller APF apparent power than that of PHC, CST1 and OPT in this case, seen Table 3.5. The CST1 strategy leads to a larger APF apparent power than the OPT strategy, which has the optimal APF apparent power when the recommended power qualities are satisfied. Table 3.8 shows that effective powers P_e and Q_e are minimum in the OPT strategy. The desired circulating current is 0, see Fig. 3.8, showing that the desired currents are tracked and the SM capacitor voltages are controlled to the desired level. Fig. 3.10 (c) shows the RMS amplitude spectrum of source currents before and after compensation in this case.

3.7.4. Case D

In this case, a single-phase voltage sag to the ground occurred in phase-a is considered in the PCC voltages (Eq. (3.97) in kV), the other two phases contain fundamental and harmonic components. The load currents are as Eq. (3.81) (KA).

$$\begin{cases} v_{sa} = 0 \\ v_{sb} = 30.4\sin\left(\omega t - \frac{2\pi}{3}\right) + 5\sin\left(5\omega t + \frac{2\pi}{3}\right) + 2\sin\left(7\omega t - \frac{2\pi}{3}\right) \\ v_{sc} = 30.4\sin\left(\omega t + \frac{2\pi}{3}\right) + 5\sin\left(5\omega t - \frac{2\pi}{3}\right) + 2\sin\left(7\omega t + \frac{2\pi}{3}\right) \end{cases} \tag{3.97}$$

Total harmonic distortion in v_{sb} and v_{sc} is 17.71%. The zero and negative sequence voltage unbalance factor are $V_{s01}/V_{s+1} = 50\%$ and $V_{s-1}/V_{s+1} = 50\%$. Table 3.6 gives the PHC and UPF strategy with the circulating current of only 1st order, the power factor of the UPF strategy is less than 1 due to the zero sequence component in PCC voltages. Under the OPT strategy, the desired source currents are not harmonic free and with slight imbalance (the source current unbalance factor is 0.8%), the circulating current contain the 1st, 5th and 7th frequency component.

Similar with the OPT strategy, the circulating current in CST1-CST3 containing 1st, 5th and 7th orders. Further, the source currents of CST1 have the same total harmonic distortion as OPT but 5th and 7th individual harmonic distortion of the CST1 source currents are respectively 3.0% and 4.0%, different with those in the

OPT strategy. The CST2 and CST3 source currents have the same 5th and 7th IHD as OPT, the CUF in the CST2 and CST3 source currents is respectively 0% and 5%, different with the CUF in the OPT strategy. The circulating current in CST4 is of 1st frequency order while the OPT circulating current contains 1st, 5th and 7th frequency orders. It can be seen from Table 3.6 that all the other strategies lead to a larger APF apparent power than the OPT strategy. The powers P_e and Q_e in Table 3.8 directly show the power consumption of APF under the OPT strategy less than others. Fig. 3.9 shows the simulation results of the OPT strategy under Case D. The spectrum of source currents of various strategies have been given in Fig. 3.10 (d).

Table 3.6.: Result summary of Case D

case		D						
strategy		PHC	UPF	CST1	CST2	CST3	CST4	**OPT**
5th IHD (%)	a	0	16.45	3.0	4.0	4.0	4.0	**4.0**
	b	0	16.45	3.0	4.0	4.0	4.0	**4.0**
	c	0	16.45	3.0	4.0	4.0	4.0	**4.0**
7th IHD (%)	a	0	6.58	4.0	3.0	3.0	3.0	**3.0**
	b	0	6.58	4.0	3.0	3.0	3.0	**3.0**
	c	0	6.58	4.0	3.0	3.0	3.0	**3.0**
THD_j (%)	a	0	17.71	5.0	5.0	5.0	5.0	**5.0**
	b	0	17.71	5.0	5.0	5.0	5.0	**5.0**
	c	0	17.71	5.0	5.0	5.0	5.0	**5.0**
CUF (%)		0	50	0.8	0	5.0	0.8	**0.8**
Circulating Current		$\{1\}$	$\{1\}$	$\{1,5,7\}$	$\{1,5,7\}$	$\{1,5,7\}$	$\{1\}$	$\mathbf{\{1,5,7\}}$
pf		0.8040	0.9129	0.8123	0.8098	0.8291	0.8129	**0.8129**
I_e (A)		141.548	228.589	127.790	127.038	128.639	127.552	**126.980**
$p_{L,dc}$ (MW)		31.199	31.199	31.199	31.199	31.199	31.199	**31.199**
S_{APF} (MVA)		11.968	19.327	10.805	10.741	10.881	10.785	**10.736**

3.8. Discussion

In this chapter, an optimal current operation strategy for a delta-connected CHB-based shunt APF under non-ideal grid conditions is presented that minimizes the APF apparent power and satisfies requirements on average power balance, power factor constraint, source current distortion constraint as per IEEE STD-519 and imbalance characteristics as per IEEE STD-1159. The design approach is explained step by step including the appropriate analysis to formulate an optimization problem, the proper treatment of the non-convex optimization based on local linearisation and iterative sequential programming. The proposed algorithm converges to a local optimal solution mostly within 2 or 3 iterations based on our computer simulations which represent typical scenarios of non-ideal grid voltage and load current conditions. The APF apparent power can be further reduced when the constraints are more relaxed. Furthermore, this proposed current strategy can be extended for other objectives, for example, branch current selective harmonic minimization, i.e., minimization of I_{en} (n is the frequency order of concern) in Eq. (3.31) subjected to constraints as Eqs. (3.57b)-(3.57g). It can be expectd that the resulted problem is still a nonconvex problem and the introduced sequential convex programming can be used to obtain the optimal solution. The strategy gives insight into the apparent power minimization (or other objectives minimization/maximization) subjected to constraints for some other converter topologies which also let circulating current(s) flow inside, such as MMCs.

The optimal current strategy is proposed when power systems contain harmonics and/or unbalance. The simulations in Section 3.7 have been implemented with various PCC voltages in steady state (or continuous state). In fact, the proposed current strategy can also be applied when some types of disturbances (temporarily or permanently) occur, such as the PCC voltage dips as Fig. 2.9.(e) or swell, as long as the duration of the disturbances is more than one fundamental period and the harmonics contained are low orders. It should be noticed that under two situations it may be impractical to use the proposed current strategy:

- The frequencies in the PCC voltage disturbances appear as a wide band spectrum instead of discrete values, such as Fig. 2.9.(a).

- The PCC voltage disturbances with short duration (less than one fundamental period) occur, such as Fig. 2.9.(c), which is inherently a high-frequency phenomenon up to low gigahertz range.

The foundation to establish the proposed current strategy has been lost under these two situations since it may be impractical to see dq values as constants by using dq transformations.

According to IEC 61000-4-7, it is possible to group interharmonics into harmonics, which uses a method based on the concept of grouping shown in Fig. 3.5. It supposes that the interharmonics are attributed to the harmonics. Note that when processing signals according to this standard, the frequency resolution of the spectrum is 5 Hz for both 50 Hz and 60 Hz power systems. Take the 3rd harmonic 150 Hz in Fig. 3.5 as an example, this standard supposes that the frequencies between 125 Hz-175 Hz can be grouped to the frequency 150 Hz. If the concept of grouping is adopted, it may be possible to use the proposed current operation strategy. The harmonic group magnitude calculation and more analysis result can be found in the standard.

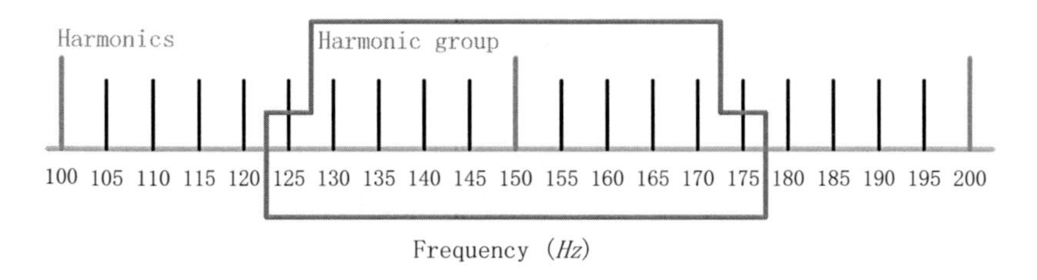

Figure 3.5.: Harmonic group according to IEC standard 61000-4-7

Table 3.7.: APF power components of various strategies with unit of **MW** and **MVar** for active and reactive Power (Case A-B)

	Case A				Case B			
	PHC	UPF	CST1	**OPT**	PHC	UPF	CST1	**OPT**
$p*_{APF,dc}$	0	0	0	**0**	0	0	0	**0**
$q*_{APF,dc}$	-7.918	-7.918	-7.918	$\mathbf{-7.918}$	-5.939	-5.939	-5.939	$\mathbf{-5.939}$
P_e	3.284	3.284	2.928	**2.857**	2.653	3.728	2.345	**2.291**
Q_e	4.871	4.871	4.062	**4.045**	3.952	5.326	3.357	**3.341**
P_2	1.866	1.866	1.866	**1.866**	1.438	2.988	1.438	**1.438**
Q_2	1.866	1.866	1.866	**1.866**	1.837	4.015	1.837	**1.837**
P_4	2.150	2.150	1.832	**1.726**	1.699	1.699	1.448	**1.365**
Q_4	2.150	2.150	1.832	**1.726**	1.699	1.699	1.449	**1.364**
P_6	0.645	0.645	0.751	**0.539**	0.815	0.815	0.775	**0.661**
Q_6	3.654	3.654	2.913	**2.913**	2.818	2.818	2.250	**2.246**
P_8	1.505	1.505	1.081	**1.187**	1.190	1.190	0.855	**0.939**
Q_8	1.505	1.505	1.081	**1.187**	1.190	1.190	0.855	**0.939**

Table 3.8.: APF power components of various strategies with unit of **MW** and **MVar** for active and reacitve power (Case C-D)

	Case C				Case D			
	PHC	UPF	CST1	**OPT**	PHC	UPF	CST1	**OPT**
$p*_{APF,dc}$	0	0	0	**0**	0	0	0	**0**
$q*_{APF,dc}$	-7.731	-7.731	-7.731	**−7.731**	-5.154	-5.154	-5.154	**−5.154**
P_e	3.387	2.205	3.000	**2.935**	2.742	3.602	2.427	**2.385**
Q_e	4.959	2.433	4.102	**4.085**	4.062	5.504	3.487	**3.475**
P_2	2.025	1.866	1.960	**1.960**	1.621	3.291	1.536	**1.536**
Q_2	2.025	1.866	1.960	**1.960**	2.371	5.025	2.275	**2.276**
P_4	2.045	0.346	1.730	**1.623**	1.543	1.062	1.372	**1.297**
Q_4	2.045	0.346	1.730	**1.623**	1.538	1.592	1.361	**1.286**
P_6	0.802	0.692	0.892	**0.731**	1.057	0.590	0.946	**0.874**
Q_6	3.729	1.246	2.951	**2.949**	2.673	1.156	2.082	**2.078**
P_8	1.503	0.861	1.090	**1.194**	1.114	0.801	0.809	**0.886**
Q_8	1.503	0.861	1.090	**1.194**	1.133	1.070	0.858	**0.932**
P_{10}	0.354	0.085	0.301	**0.283**	0.264	0.105	0.230	**0.217**
Q_{10}	0.354	0.085	0.301	**0.283**	0.264	0.152	0.224	**0.211**
P_{12}	0.389	0.166	0.297	**0.308**	0.276	0.115	0.211	**0.219**
Q_{12}	0.106	0.106	0.057	**0.081**	0.118	0.077	0.081	**0.093**
P_{14}	0.099	0.054	0.071	**0.078**	0.074	0.043	0.053	**0.058**
Q_{14}	0.099	0.054	0.071	**0.078**	0.074	0.054	0.054	**0.059**

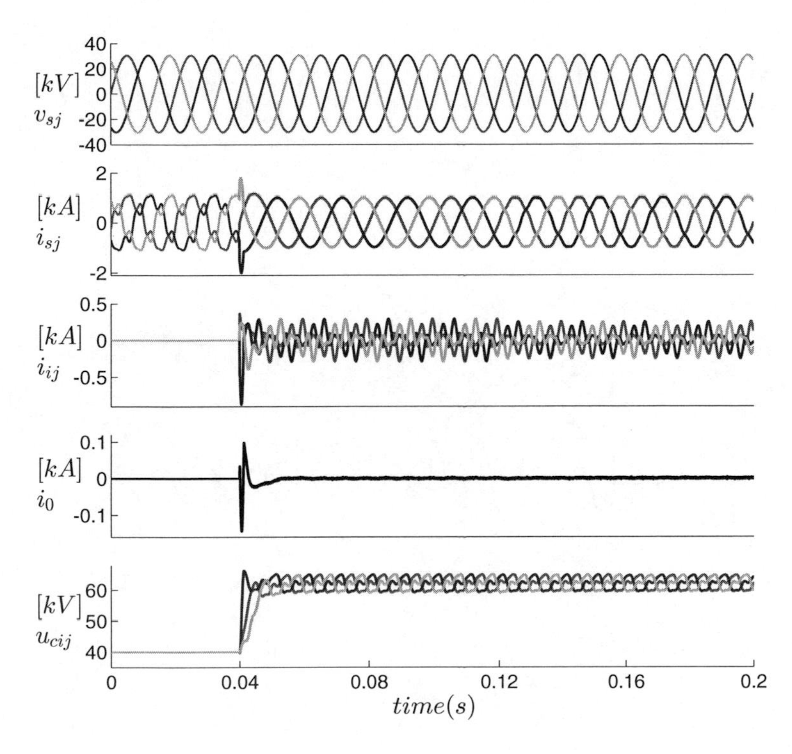

Figure 3.6.: Case A with PHC (0.04 s-0.08 s), UPF (0.08 s-0.12 s), CST1 (0.12 s-0.16 s), and OPT (0.16 s-0.2 s). The APF injection at $t = 0.04$ s. From top to bottom are PCC voltages, source currents, branch currents, circulating current, the sum of SM capacitor voltages of each branch.

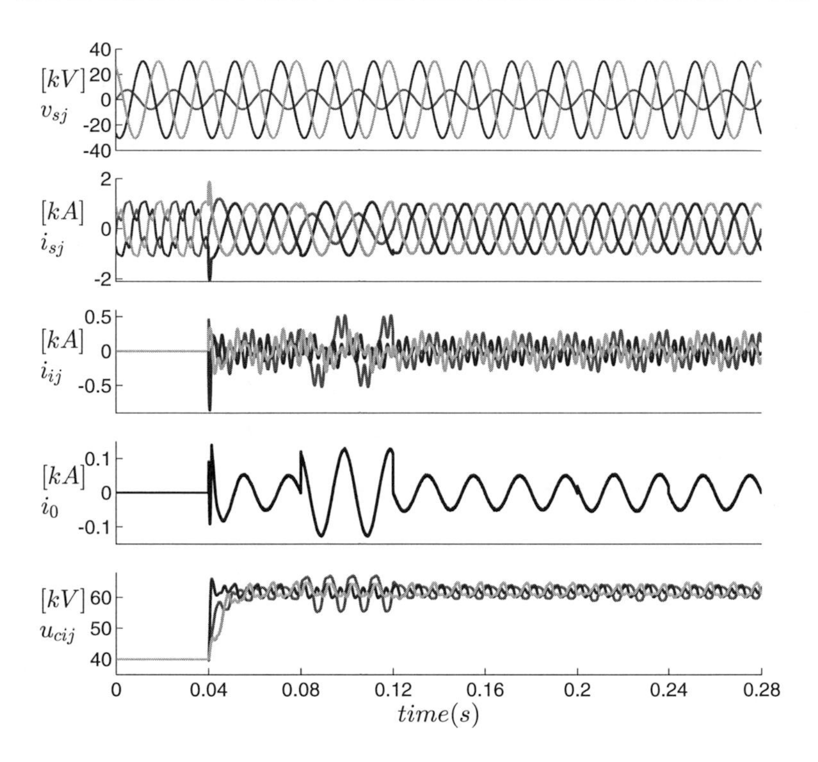

Figure 3.7.: Case B with PHC (0.04 s-0.08 s), UPF (0.08 s-0.12 s), CST1 (0.12 s-0.16 s), CST2 (0.16 s-0.2 s), CST3 (0.2 s-0.24 s), and OPT (0.24 s-0.28 s). The APF injection at $t = 0.04$ s. From top to bottom are PCC voltages, source currents, branch currents, circulating current, the sum of SM capacitor voltages of each branch.

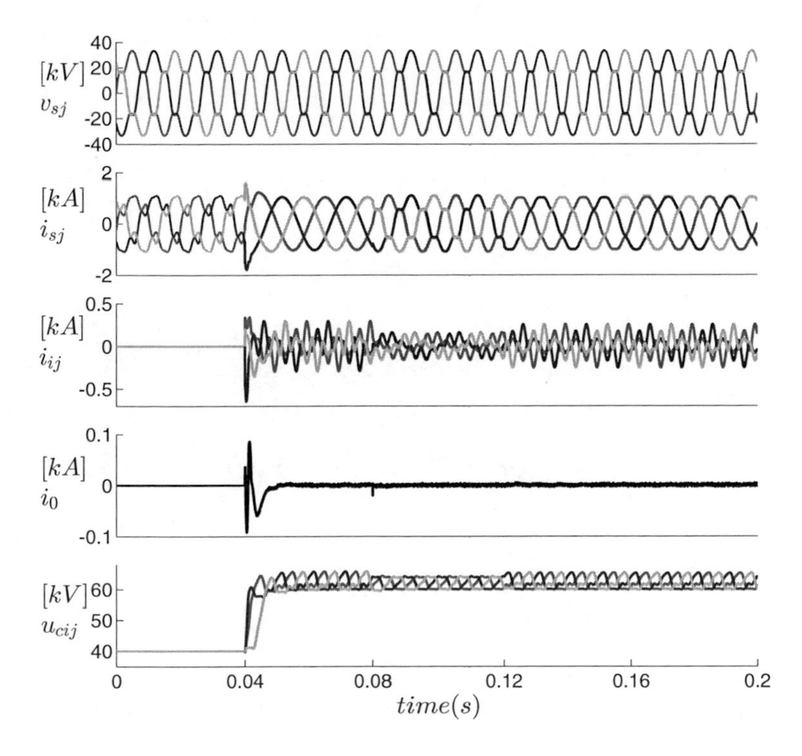

Figure 3.8.: Case C with PHC (0.04 s-0.08 s), UPF (0.08 s-0.12 s), CST1 (0.12 s-0.16 s) and OPT (0.16 s-0.2 s). The APF injection at $t = 0.04$ s. From top to bottom are PCC voltages, source currents, branch currents, circulating current, the sum of SM capacitor voltages of each branch.

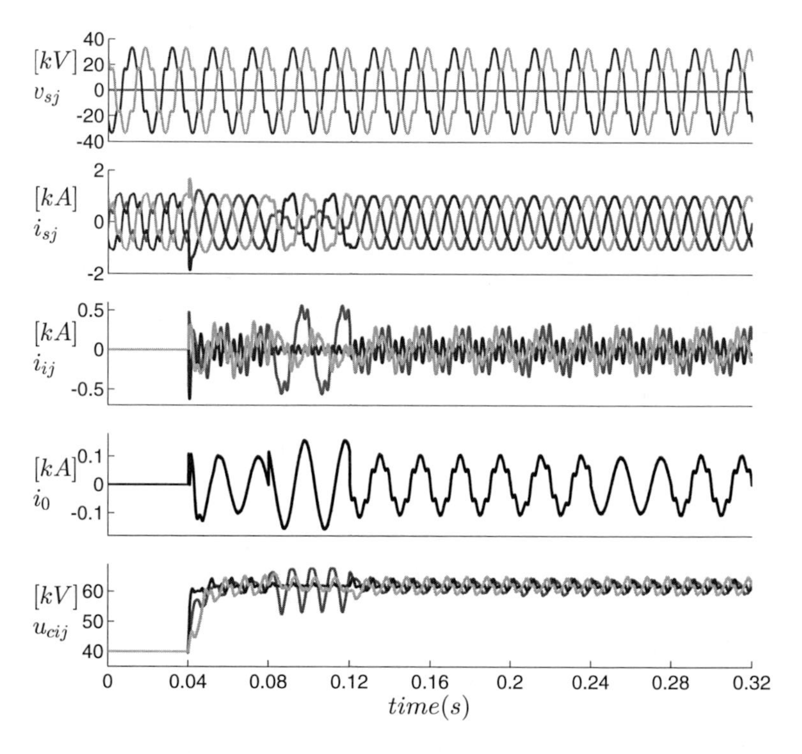

Figure 3.9.: Case D with with PHC (0.04 s-0.08 s), UPF (0.08 s-0.12 s), CST1 (0.12 s-0.16 s), CST2 (0.16 s-0.2 s), CST3 (0.2 s-0.24 s), CST4 (0.24 s-0.28 s), and OPT (0.28 s-0.32 s). The APF injection at $t = 0.04\,\mathrm{s}$. From top to bottom are PCC voltages, source currents, branch currents, circulating current, the sum of SM capacitor voltages of each branch.

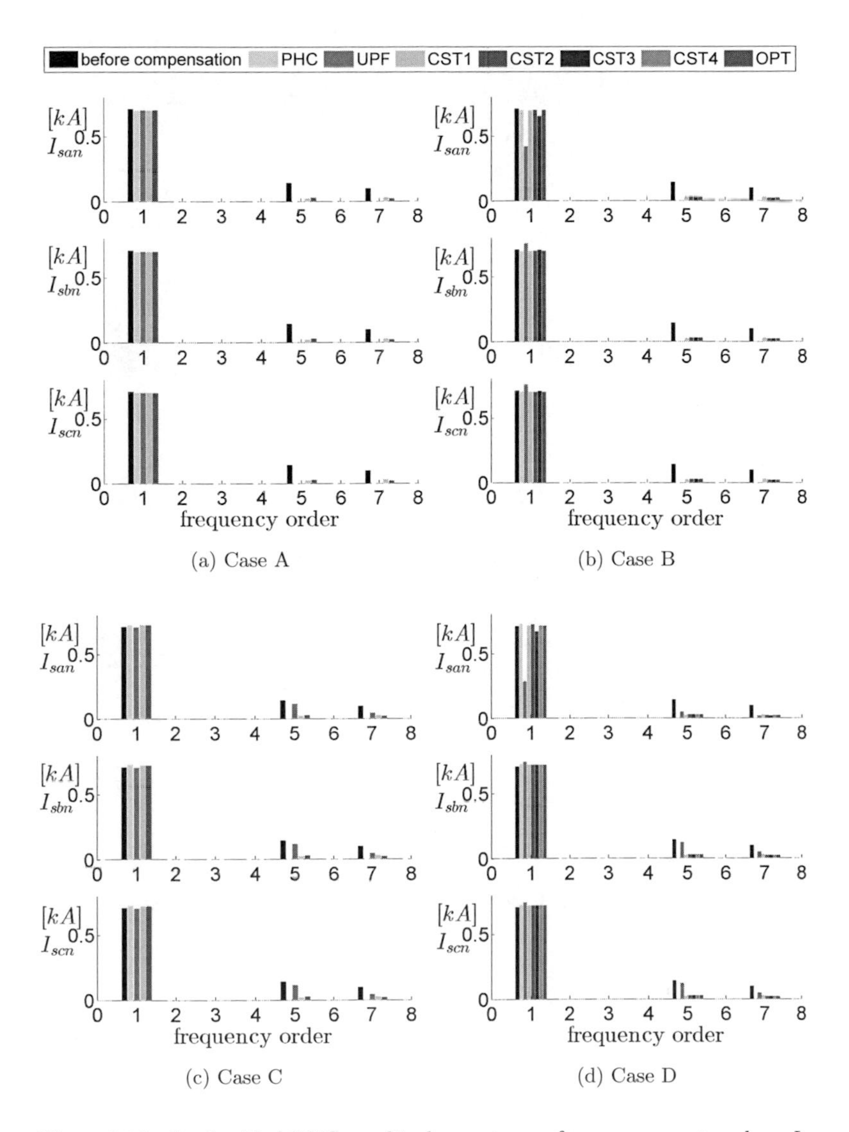

Figure 3.10.: Single-sided RMS amplitude spectrum of source currents, where I_{sjn} represents the RMS value of the nth order phase-j source current.

4. Harmonic Interaction Analysis

This chapter presents harmonic interaction analysis based on the derivation of a time-invariant state-space model integrating branch currents and branch energies as state variables and switching functions (modulation signals) as control inputs which are unknown beforehand. Based on the harmonic interaction analysis, this chapter is capable of predicting how the harmonics propagating through the system and providing precise calculation of electrical quantities and non-electrical quantities (switching functions/modulation signals) which contain multiple frequency orders with symmetrical components, giving an accurate reference for understanding, designing, and controlling of delta-connected CHB-based shunt APFs.

4.1. Necessity of Harmonic Interaction Analysis

The last chapters have introduced that each branch of the delta-connected CHB is comprising a number of modular H-bridge submodules (SMs) fed by dc-link capacitors. The SM dc-link application with small electrolytic capacitors or even film capacitors leads to higher amplitude low-frequency (<2 kHz) capacitor voltage ripple, however, can achieve lower cost and volumn. In this way, one of the major problems that should be solved is harmonic interaction, see Fig. 4.1: each APF branch voltage results from the interaction between the SM capacitor sum voltage and the switching function in the corresponding branch, at the same time, the branch power flow is the product of the branch current and voltage, leading to significant low-frequency harmonics in SM capacitor voltages. It is significant to develop the harmonic interaction analysis so as to predict how harmonics propagate through the system and quantify the electrical and non-electrical variables, which are essential in the APF designing and controlling stages. At the APF

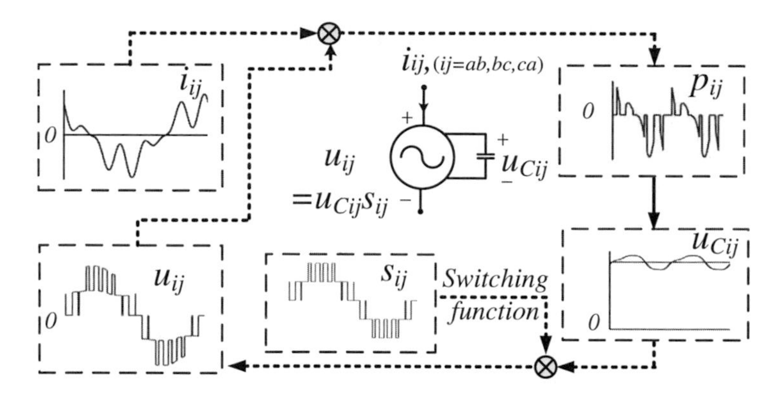

Figure 4.1.: Interaction of various quantities in one APF branch

designing stage such analysis is helpful for circuit parameters design and semiconductor devices selection. At the controlling stage, admittedly, the harmonic voltages/currents/powers can be properly tuned by controllers, various control strategies have been proposed in [FHA15, GDM15, EVMMRZ07, CLC12, AROARd+15], however, for some control strategies that are sensitive to model errors, without accurate variable quantification, time-consuming simulations have to be performed in order to find a suitable combination of steady control inputs and state variables, resulting in poor dynamic response.

Fig. 4.2 shows the comparison of the ac-side dynamics when the harmonic interaction is ignored (Fig. 4.2. (a)) and considerd (Fig. 4.2. (b)). In the state-space model for the ac-side dynamic, $\boldsymbol{x}$ is the state variable vector collecting three-phase branch currents and $\boldsymbol{u}$ is the control input vector collecting three-phase switching signals. Conventionally SM capacitor voltages are modelled time-invariant, based on which the branch currents are represented in state-space with time-invariant system, input and disturbance matrix, see Fig. 4.2. (a). In this linear model the operating switching signals are calculated as the ratio between the operating branch voltages and the average capacitor sum voltages, indicating that switching signals of one frequency order can only invoke branch voltages/currents with the same frequency order. This linear model requires that the capacitor voltage ripple is little, which happens with high average capacitor voltage or high capacitance. When reduced capacitance is applied for decreased system bulk, dimension and expenses, along with a consequence of large capacitor voltage variation, the ac-side dynamics cannot be approximated by linear equations, as shown in Fig. 4.2.(b), in which the capacitor voltage harmonics are not ignored. In this description the operating switching signals are the ratio between the operating branch voltages and the instantaneous capacitor sum voltages, indicating that switching signals of one frequency order can invoke branch voltages/currents not only with the same frequency order but also with other frequency orders due to the coupling of capacitor voltage harmonics and switching signals.

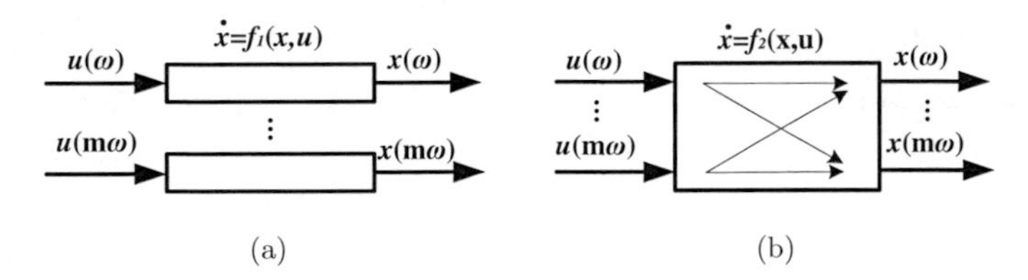

Figure 4.2.: Comparison of (a) linear system and (b) nonlinear system

This chapter aims at harmonic interaction analysis of the delta-connected CHB as shunt APF via the derivation of a time-invariant model in multiple dq-frame from the time-varying abc-frame model integrating the ac- and dc-side dynamics. In the derived time-invariant model variables associated with different frequency orders in sequences can be decoupled. Based on the analysis, the detailed expression of harmonic sequences in power flows is given for the first time. A complete relation between power flows and SM capacitor voltages is given, leading to an accurate SM capacitor voltage calculation. Via the harmonic interaction analysis, the branch voltage dynamic is described as the coupling of the capacitor voltage with harmonics and the switching signal, resulting in the accurate switching signal calculation. One advantage of the proposed approach is that it is more accurate than the conventional analysis in the present research. Another advantage is that the sequence quantities can be decoupled. The proposed approach can also be useful for other power-electronic-based systems, especially for the applications where the low-frequency harmonics in the capacitor voltage should be an important consideration, such as ac motor drives where the overheating of components and mechanical oscillation are a concern, communication systems which should meet certain noise requirement or applications where the converter is connected to the electrical grid which should satisfy certain grid code.

4.2. Time-varying Model and Properties

This section starts with variable definitions in the delta-connected CHB-based shunt APF system and an average model integrating the ac- and dc-side dynamics in *abc*-frame is built. Afterwards the properties of the variables in the model are presented.

In this chapter the object of study is the APF branch variables, also as phase-phase variables, with the definition of vector $\boldsymbol{x}^{abc}$ for the phase-phase variables as Eq. (4.1). Eq. (4.2) represents the dc component in $\boldsymbol{x}^{abc}$; Eq. (4.3) is a vector with the nth frequency component of the three-phase variables; Eqs. (4.4)-(4.6) show the nth positive-, negative-, and zero-sequence component respectively.

$$\boldsymbol{x}^{abc} = \begin{bmatrix} x_{ab} & x_{bc} & x_{ca} \end{bmatrix}^\top \tag{4.1}$$

$$\boldsymbol{x}^{abc}_{dc} = \begin{bmatrix} x_{ab,dc} & x_{bc,dc} & x_{ca,dc} \end{bmatrix}^\top \tag{4.2}$$

$$\boldsymbol{x}^{abcn} = \begin{bmatrix} x_{abn} & x_{bcn} & x_{can} \end{bmatrix}^\top \tag{4.3}$$

$$\boldsymbol{x}^{abc+n} = \begin{bmatrix} x_{ab+n} & x_{bc+n} & x_{ca+n} \end{bmatrix}^\top \tag{4.4}$$

$$\boldsymbol{x}^{abc-n} = \begin{bmatrix} x_{ab-n} & x_{bc-n} & x_{ca-n} \end{bmatrix}^\top \tag{4.5}$$

$$\boldsymbol{x}^{abc0n} = \begin{bmatrix} x_{ab0n} & x_{bc0n} & x_{ca0n} \end{bmatrix}^\top \tag{4.6}$$

the superscript *abc* represents *abc*-frame, $\top$ indicates transpose.

Dynamics of three-phase branch currents: When a continuous time average model is used, the piece-wise feature of switches is approximated with an overall smooth trait and balancing between the SM capacitor voltages is assumed, the SMs are replaced by an equivalent voltage source, whose output voltage is the product of the submodule capacitor sum voltages as introduced in Eq. (3.12) and the modulation signal/switching function, see Eq. (4.8), where the modulation signal s_{ij}, $(ij = ab, bc, ca)$ is now a continuous variable within the range of $[-1, 1]$. With this basis, the dynamic of the three-phase branch currents can be described by the following average model in abc-frame.

$$\frac{d}{dt}\boldsymbol{i}^{abc} = -\frac{r}{L}\boldsymbol{i}^{abc} - \frac{1}{L}\boldsymbol{u}^{abc} + \frac{1}{L}\boldsymbol{v}_s^{abc} \tag{4.7}$$

$$\boldsymbol{u}^{abc} = \boldsymbol{u}_C^{abc} \circ \boldsymbol{s}^{abc} = \begin{bmatrix} u_{Cab}s_{ab} & u_{Cbc}s_{bc} & u_{Cca}s_{ca} \end{bmatrix}^\top \tag{4.8}$$

where $\boldsymbol{i}^{abc} = \begin{bmatrix} i_{ab} & i_{bc} & i_{ca} \end{bmatrix}^\top$, $\boldsymbol{u}^{abc} = \begin{bmatrix} u_{ab} & u_{bc} & u_{ca} \end{bmatrix}^\top$, $\boldsymbol{v}_s^{abc} = \begin{bmatrix} v_{sab} & v_{sbc} & v_{sca} \end{bmatrix}^\top$, and $\boldsymbol{s}^{abc} = \begin{bmatrix} s_{ab} & s_{bc} & s_{ca} \end{bmatrix}^\top$. $\circ$ represents element multiplication defined in Section 1.4.

Dynamics of three-phase SM capacitor sum energies: Define the submodule capacitor sum energies into a vector $\boldsymbol{w}^{abc} = \begin{bmatrix} w_{ab} & w_{bc} & w_{ca} \end{bmatrix}^\top$, where $w_{ab,bc,ca}$ has been introduced in Eq. (3.13), and $\boldsymbol{p}^{abc} = \begin{bmatrix} p_{ab} & p_{bc} & p_{ca} \end{bmatrix}^\top$, where $p_{ab,bc,ca}$ is the instantaneous power in the corresponding branch. The dynamic of the three-phase SM capacitor sum energies is given by

$$\frac{d}{dt}\boldsymbol{w}^{abc} = \boldsymbol{p}^{abc} \tag{4.9}$$

$$\boldsymbol{p}^{abc} = \boldsymbol{u}^{abc} \circ \boldsymbol{i}^{abc} = \begin{bmatrix} u_{ab}i_{ab} & u_{bc}i_{bc} & u_{ca}i_{ca} \end{bmatrix}^\top \tag{4.10}$$

An inspection of the model Eqs. (4.7) and (4.9) reveals a number of properties:

1. The system is time-varying. The measurable PCC line-line voltages $v_{sab,sbc,sca}$ (or calculated by measurable PCC voltages) contained in $\boldsymbol{v}_s^{abc}$ can be either ideal or non-ideal, where the ideal voltages contain only the positive-sequence fundamental frequency component and the non-ideal voltages can be unbalanced and/or distorted. The general description of $\boldsymbol{v}_s^{abc}$ is expressed by Eq. (4.11) with the highest considered frequency order $M \leq 40$th.

$$\boldsymbol{v}_s^{abc} = \sum_{n=1}^{M} \boldsymbol{v}_s^{abcn} = \sum_{n=1}^{M} (\boldsymbol{v}_s^{abc+n} + \boldsymbol{v}_s^{abc-n}) \tag{4.11}$$

The branch current $i_{ab,bc,ca}$ contained in $\boldsymbol{i}^{abc}$ (in turn, branch voltage $u_{ab,bc,ca}$ contained in $\boldsymbol{u}^{abc}$) in Eq. (4.7) contains the fundamental frequency and harmonics components so that the source currents after compensation satisfy the recommended practice for power quality. The reference branch current determination has been presented in Section 4.5. The general description of three-phase branch currents and branch voltages are as Eqs. (4.12) - (4.13). It should be noticed that Eqs. (4.12) - (4.13) represent a description at steady state.

$$\boldsymbol{i}^{abc} = \sum_{n=1}^{M} \boldsymbol{i}^{abcn} = \sum_{n=1}^{M} (\boldsymbol{i}^{abc+n} + \boldsymbol{i}^{abc-n} + \boldsymbol{i}^{abc0n}) \tag{4.12}$$

$$\boldsymbol{u}^{abc} = \sum_{n=1}^{M} \boldsymbol{u}^{abcn} = \sum_{n=1}^{M} (\boldsymbol{u}^{abc+n} + \boldsymbol{u}^{abc-n} + \boldsymbol{u}^{abc0n}) \tag{4.13}$$

The three-phase capacitor sum voltages contain a dc component as well as multiple frequency components. The generic expression of three-phase steady-state capacitor sum voltages/energies is as follows.

$$\begin{aligned} \boldsymbol{u}_C^{abc} &= \boldsymbol{u}_{C,dc}^{abc} + \sum_{n=1}^{M} \boldsymbol{u}_C^{abcn} \\ &= u_{C,dc} \begin{bmatrix} 1 & 1 & 1 \end{bmatrix}^\top + \sum_{n=1}^{M} (\boldsymbol{u}_C^{abc+n} + \boldsymbol{u}_C^{abc-n} + \boldsymbol{u}_C^{abc0n}) \end{aligned} \tag{4.14}$$

$$\begin{aligned} \boldsymbol{w}^{abc} &= \boldsymbol{w}_{dc}^{abc} + \sum_{n=1}^{M} \boldsymbol{w}^{abcn} \\ &= w_{dc} \begin{bmatrix} 1 & 1 & 1 \end{bmatrix}^\top + \sum_{n=1}^{M} \left(\boldsymbol{w}^{abc+n} + \boldsymbol{w}^{abc-n} + \boldsymbol{w}^{abc0n}\right) \end{aligned} \tag{4.15}$$

where $\boldsymbol{u}_{C,dc}^{abc} = \begin{bmatrix} u_{Cab,dc} & u_{Cbc,dc} & u_{Cca,dc} \end{bmatrix}^\top$ and $\boldsymbol{w}_{dc}^{abc} = \begin{bmatrix} w_{ab,dc} & w_{bc,dc} & w_{ca,dc} \end{bmatrix}^\top$ respectively represents the dc component in $\boldsymbol{u}_C^{abc}$ and $\boldsymbol{w}^{abc}$, satisfying $u_{Cab,dc} = u_{Cbc,dc} = u_{Cca,dc} = u_{C,dc}$ and $w_{ab,dc} = w_{bc,dc} = w_{ca,dc} = w_{dc}$.

Seen in Eq. (4.9), a dc component cannot be sustained in $\boldsymbol{p}^{abc}$ at steady state, because this dc component would lead to a drifting $w_{ab,bc,ca}$, thus at steady state $\boldsymbol{p}_{dc}^{abc} = \begin{bmatrix} p_{ab,dc} & p_{bc,dc} & p_{ca,dc} \end{bmatrix}^\top = \mathbf{0}$,

$$\boldsymbol{p}^{abc} = \boldsymbol{p}_{dc}^{abc} + \sum_{n=1}^{M} \boldsymbol{p}^{abcn} = \sum_{n=1}^{M} \left(\boldsymbol{p}^{abc+n} + \boldsymbol{p}^{abc-n} + \boldsymbol{p}^{abc0n}\right) \tag{4.16}$$

2. The system contains harmonic interaction. Eq. (4.8) shows that the branch voltage is resulted from the harmonic interaction between the SM capacitor sum voltage and the modulation signal. Eq. (4.10) shows that the instantaneous power flow is the product of the branch voltage and current in the corresponding branch.

3. The conventional approach for calculating steady-state modulation signals is given by $\boldsymbol{s}^{abc} = \boldsymbol{u}^{abc}/u_{C,dc}$ where the denominator is constant by disregarding the harmonics in the capacitor sum voltage. The conventional calculation of $\boldsymbol{s}^{abc}$ is although not accurate, still reasonable when the capacitor voltage harmonics have low-magnitude. If the harmonics in $u_{Cab,Cbc,Cca}$ have high-magnitude, it is unreasonale to disregard the capacitor voltage harmonics. Here electrolytic capacitors or reliable film capacitors with low-capacitance have been applied for submodule dc-link, which lead to high-magnitude capacitor voltage harmonics, thus the capacitor voltage harmonics should not be disregarded to calculate $\boldsymbol{s}^{abc}$.

4. Although the modulation signals are unknown beforehand, it can be still derived that the dc component cannot be sustained in the modulation signals, i.e., the steady state modulation signals contain only oscillating components. Since multiplication of the dc component in $s_{ab,bc,ca}$ and capacitor voltages in Eq. (4.8) would result in a dc component in $u_{ab,bc,ca}$ as well as in $i_{ab,bc,ca}$. And a dc component would be generated at the right side of Eq. (4.9) by the multiplication of the dc component in $u_{ab,bc,ca}$ and $\boldsymbol{i}_{ab,bc,ca}$, giving a drifting $w_{ab,bc,ca}$. Therefore the modulation signals at steady state can be expressed by

$$\boldsymbol{s}^{abc} = \sum_{n=1}^{M} \boldsymbol{s}^{abcn} = \sum_{n=1}^{M} \left(\boldsymbol{s}^{abc+n} + \boldsymbol{s}^{abc-n} + \boldsymbol{s}^{abc0n}\right) \tag{4.17}$$

5. To accurately calculate steady-state $\boldsymbol{s}^{abc}$, the quantification of harmonics in the capacitor sum voltage should be achieved. The capacitor voltage harmonic quantification is based on the relationship

$$\frac{d}{dt}\begin{bmatrix} w_{ab} \\ w_{bc} \\ w_{ca} \end{bmatrix} = \begin{bmatrix} p_{ab} \\ p_{bc} \\ p_{ca} \end{bmatrix} = C_{sum}\boldsymbol{u}_C^{abc} \circ \frac{d}{dt}\boldsymbol{u}_C^{abc} = C_{sum}\begin{bmatrix} u_{Cab} \cdot \frac{d}{dt}u_{Cab} \\ u_{Cbc} \cdot \frac{d}{dt}u_{Cbc} \\ u_{Cca} \cdot \frac{d}{dt}u_{Cca} \end{bmatrix} \tag{4.18}$$

The conventional approach for capacitor voltage harmonic quantification is based on the approximation of Eq. (4.18) to the following

$$\boldsymbol{p}^{abc} \approx C_{sum}u_{C,dc} \cdot \frac{d}{dt}\boldsymbol{u}_C^{abc} \tag{4.19}$$

The conventional calculation of capacitor voltages in each branch is then as following by use of Eq. (4.8),

$$\begin{aligned}\boldsymbol{u}_C^{abc} &\approx u_{C,dc} + \frac{1}{C_{sum}u_{C,dc}}\int \boldsymbol{p}^{abc}dt \\ &= u_{C,dc} + \frac{1}{C_{sum}u_{C,dc}}\int \begin{bmatrix} u_{ab}i_{ab} & u_{bc}i_{bc} & u_{ca}i_{ca}\end{bmatrix}^\top dt \end{aligned} \tag{4.20}$$

The approximation of Eq. (4.18) to Eq. (4.19) is reasonable if the capacitor voltage harmonics have low-amplitudes. The approximation leads to unaccurate quantification of capacitor voltage harmonics if they have high-amplitudes. The quantification of capacitor voltage harmonics presented in Section 4.5.2 is based on Eq. (4.18) without approximation.

4.3. Observations of Power Flow and Branch Voltage Model

The dynamics of branch currents and energies (Eqs. (4.7) and (4.9)) show that the coupling mainly lies in Eqs. (4.8) and (4.10). An observation of Eqs. (4.8) and (4.10) will be given to show some properties of the interaction.

4.3.1. Foundation

In order to derive a time invariant representation for Eqs. (4.7) and (4.9), the variables should be expressed in a way that the symmetrical components in each frequency order can be seperated and transformed into their corresponding synchrounous frame while the coupling with other variables with harmonic sequences is reserved.

Define the following vectors with italic bold letters $\boldsymbol{x}^{dq+n} = \begin{bmatrix} x_{d+n} & x_{q+n}\end{bmatrix}^\top$, $\boldsymbol{x}^{dq-n} = \begin{bmatrix} x_{d-n} & x_{q-n}\end{bmatrix}^\top$, $\boldsymbol{x}^{dq0n} = \begin{bmatrix} x_{d0n} & x_{q0n}\end{bmatrix}^\top$, Eq. (4.21) shows that each frequency in the oscillating components of variables can be seperated to positive-, negative- and zero-sequence, and each sequence can be described with equivelent dq variables.

$$\boldsymbol{x}^{abcn} = \boldsymbol{x}^{abc+n} + \boldsymbol{x}^{abc-n} + \boldsymbol{x}^{abc0n} \tag{4.21}$$

$$= \boldsymbol{T}_{+n}^{2r/3s}\boldsymbol{x}^{dq+n} + \boldsymbol{T}_{-n}^{2r/3s}\boldsymbol{x}^{dq-n} + \boldsymbol{T}_{0n}^{2r/3s}\boldsymbol{x}^{dq0n}$$

where $\boldsymbol{T}_{+n}^{2r/3s}$, $\boldsymbol{T}_{-n}^{2r/3s}$, and $\boldsymbol{T}_{0n}^{2r/3s}$ are the same as in Eq. (3.18).

$$+n\omega : \boldsymbol{x}^{abc+n} = \boldsymbol{T}_{+n}^{2r/3s}\boldsymbol{x}^{dq+n} \tag{4.22a}$$

$$-n\omega : \boldsymbol{x}^{abc-n} = \boldsymbol{T}_{-n}^{2r/3s}\boldsymbol{x}^{dq-n} \tag{4.22b}$$

$$0n\omega : \boldsymbol{x}^{abc0n} = \boldsymbol{T}_{0n}^{2r/3s}\boldsymbol{x}^{dq0n} \tag{4.22c}$$

Define

$$\boldsymbol{x}^{dqn} = \begin{bmatrix} \boldsymbol{x}^{dq+n\top} & \boldsymbol{x}^{dq-n\top} & \boldsymbol{x}^{dq0n\top} \end{bmatrix}^\top \tag{4.23}$$

The dq variables of all the harmonics are considered and can be combined into a vector as

$$\boldsymbol{x}^{dq} = \begin{bmatrix} \boldsymbol{x}^{dq1\top} & \boldsymbol{x}^{dq2\top} \cdots & \boldsymbol{x}^{dqM\top} \end{bmatrix}^\top \tag{4.24}$$

4.3.2. Observations of Power Flow Model

As mentioned, $\boldsymbol{u}^{abc}$ and $\boldsymbol{i}^{abc}$ are of multiple frequency orders, satisfying Eqs. (4.13) and (4.12), Eq. (4.10) then can be rewritten as

$$\boldsymbol{p}^{abc} = \begin{bmatrix} p_{ab} \\ p_{bc} \\ p_{ca} \end{bmatrix} = \begin{bmatrix} u_{ab}i_{ab} \\ u_{bc}i_{bc} \\ u_{ca}i_{ca} \end{bmatrix} = \begin{bmatrix} \sum_{m=1}^{M} u_{abm} \sum_{k=1}^{M} i_{abk} \\ \sum_{m=1}^{M} u_{bcm} \sum_{k=1}^{M} i_{bck} \\ \sum_{m=1}^{M} u_{cam} \sum_{k=1}^{M} i_{cak} \end{bmatrix} \tag{4.25}$$

The following vector $\breve{\boldsymbol{p}}^{abc}$ is introduced, which represents the product of the mth order branch voltage and the kth order branch current, with $\forall m, k \in \{1, 2, \cdots, M\}$,

$$\begin{aligned} \breve{\boldsymbol{p}}^{abc} &= \begin{bmatrix} \breve{p}_{ab} & \breve{p}_{bc} & \breve{p}_{ca} \end{bmatrix}^\top \\ &= \boldsymbol{u}^{abcm} \circ \boldsymbol{i}^{abck} = \begin{bmatrix} u_{abm}i_{abk} & u_{bcm}i_{bck} & u_{cam}i_{cak} \end{bmatrix}^\top \end{aligned} \tag{4.26}$$

$$\begin{aligned} \boldsymbol{u}^{abcm} &= \boldsymbol{u}^{abc+m} + \boldsymbol{u}^{abc-m} + \boldsymbol{u}^{abc0m} \\ &= \boldsymbol{T}_{+m}^{2r/3s}\boldsymbol{u}^{dq+m} + \boldsymbol{T}_{-m}^{2r/3s}\boldsymbol{u}^{dq-m} + \boldsymbol{T}_{0m}^{2r/3s}\boldsymbol{u}^{dq0m} \end{aligned} \tag{4.27}$$

$$\begin{aligned} \boldsymbol{i}^{abck} &= \boldsymbol{i}^{abc+k} + \boldsymbol{i}^{abc-k} + \boldsymbol{i}^{abc0k} \\ &= \boldsymbol{T}_{+k}^{2r/3s}\boldsymbol{i}^{dq+k} + \boldsymbol{T}_{-k}^{2r/3s}\boldsymbol{i}^{dq-k} + \boldsymbol{T}_{0k}^{2r/3s}\boldsymbol{i}^{dq0k} \end{aligned} \tag{4.28}$$

The detailed calculation in the Appendix as Section A.2 yields the following three cases dependent on $m > k$, $m < k$, or $m = k$. The tables in Section A.3 list the resulted harmonic sequences.

- Case $m > k$, $\breve{\boldsymbol{p}}^{abc}$ contains the $(m+k)$th and $(m-k)$th order frequency component.
- Case $m < k$, $\breve{\boldsymbol{p}}^{abc}$ contains the $(m+k)$th and $(k-m)$th order frequency component.
- Case $m = k$, $\breve{\boldsymbol{p}}^{abc}$ contains the $(m+k)$th and a dc component.

(1) Case $m > k$:

$\breve{\boldsymbol{p}}^{abc}$ is consisted of the $(m+k)$th and the $(m-k)$th orders, both frequency orders contain the positive-, negative- and zero-sequence component, see Eq. (4.29). The elements in $\breve{\boldsymbol{p}}^{dq\pm(m+k)}$ and $\breve{\boldsymbol{p}}^{dq0(m+k)}$ are in Eq. (4.30). $\breve{\boldsymbol{p}}^{dq\pm(m-k)}$ and $\breve{\boldsymbol{p}}^{dq0(m-k)}$ can be seen in Eq. (4.31). Eqs. (4.30)-(4.31) provide insight to the harmonic sequence interaction. Take $m = 3$ and $k = 1$ in Eq. (4.30) as an example, the interaction of the 03th voltage and the +1th current, the interaction of the −3th voltage and the −1th current, the interaction of the +3th voltage and the 01th current will generate the +4th power.

$$\begin{aligned}
\breve{\boldsymbol{p}}^{abc} &= \breve{\boldsymbol{p}}^{abc+(m+k)} + \breve{\boldsymbol{p}}^{abc-(m+k)} + \breve{\boldsymbol{p}}^{abc0(m+k)} \\
&+ \breve{\boldsymbol{p}}^{abc+(m-k)} + \breve{\boldsymbol{p}}^{abc-(m-k)} + \breve{\boldsymbol{p}}^{abc0(m-k)} \\
&= \boldsymbol{T}^{2r/3s}_{+(m+k)}\breve{\boldsymbol{p}}^{dq+(m+k)} + \boldsymbol{T}^{2r/3s}_{-(m+k)}\breve{\boldsymbol{p}}^{dq-(m+k)} + \boldsymbol{T}^{2r/3s}_{0(m+k)}\breve{\boldsymbol{p}}^{dq0(m+k)} \\
&+ \boldsymbol{T}^{2r/3s}_{+(m-k)}\breve{\boldsymbol{p}}^{dq+(m-k)} + \boldsymbol{T}^{2r/3s}_{-(m-k)}\breve{\boldsymbol{p}}^{dq-(m-k)} + \boldsymbol{T}^{2r/3s}_{0(m-k)}\breve{\boldsymbol{p}}^{dq0(m-k)}
\end{aligned} \tag{4.29}$$

where

$$\underbrace{\begin{bmatrix} \breve{\boldsymbol{p}}^{dq+(m+k)} \\ \breve{\boldsymbol{p}}^{dq-(m+k)} \\ \breve{\boldsymbol{p}}^{dq0(m+k)} \end{bmatrix}}_{\breve{\boldsymbol{p}}^{dq(m+k)}} = \underbrace{\begin{bmatrix} \frac{1}{2}\boldsymbol{F}_1(\boldsymbol{u}^{dq0m}) & \frac{1}{\sqrt{6}}\boldsymbol{F}_1(\boldsymbol{u}^{dq-m}) & \frac{1}{2}\boldsymbol{F}_1(\boldsymbol{u}^{dq+m}) \\ \frac{1}{\sqrt{6}}\boldsymbol{F}_1(\boldsymbol{u}^{dq+m}) & \frac{1}{2}\boldsymbol{F}_1(\boldsymbol{u}^{dq0m}) & \frac{1}{2}\boldsymbol{F}_1(\boldsymbol{u}^{dq-m}) \\ \frac{1}{3}\boldsymbol{F}_1(\boldsymbol{u}^{dq-m}) & \frac{1}{3}\boldsymbol{F}_1(\boldsymbol{u}^{dq+m}) & \frac{1}{2}\boldsymbol{F}_1(\boldsymbol{u}^{dq0m}) \end{bmatrix}}_{\boldsymbol{\mathcal{F}}_1(\boldsymbol{u}^{dqm})} \underbrace{\begin{bmatrix} \boldsymbol{i}^{dq+k} \\ \boldsymbol{i}^{dq-k} \\ \boldsymbol{i}^{dq0k} \end{bmatrix}}_{\boldsymbol{i}^{dqk}}$$

$$\text{with} \quad (m > k; m < k; m = k; 0 < m, k \leq M) \tag{4.30}$$

$$\underbrace{\begin{bmatrix} \breve{\boldsymbol{p}}^{dq+(m-k)} \\ \breve{\boldsymbol{p}}^{dq-(m-k)} \\ \breve{\boldsymbol{p}}^{dq0(m-k)} \end{bmatrix}}_{\breve{\boldsymbol{p}}^{dq(m-k)}} = \underbrace{\begin{bmatrix} \frac{1}{\sqrt{6}}\boldsymbol{F}_2(\boldsymbol{u}^{dq-m}) & \frac{1}{2}\boldsymbol{F}_2(\boldsymbol{u}^{dq0m}) & \frac{1}{2}\boldsymbol{F}_2(\boldsymbol{u}^{dq+m}) \\ \frac{1}{2}\boldsymbol{F}_2(\boldsymbol{u}^{dq0m}) & \frac{1}{\sqrt{6}}\boldsymbol{F}_2(\boldsymbol{u}^{dq+m}) & \frac{1}{2}\boldsymbol{F}_2(\boldsymbol{u}^{dq-m}) \\ \frac{1}{3}\boldsymbol{F}_2(\boldsymbol{u}^{dq+m}) & \frac{1}{3}\boldsymbol{F}_2(\boldsymbol{u}^{dq-m}) & \frac{1}{2}\boldsymbol{F}_2(\boldsymbol{u}^{dq0m}) \end{bmatrix}}_{\boldsymbol{\mathcal{F}}_2(\boldsymbol{u}^{dqm})} \underbrace{\begin{bmatrix} \boldsymbol{i}^{dq+k} \\ \boldsymbol{i}^{dq-k} \\ \boldsymbol{i}^{dq0k} \end{bmatrix}}_{\boldsymbol{i}^{dqk}}$$

$$\text{with} \quad (m > k; 0 < m, k \leq M) \tag{4.31}$$

(2) Case $m < k$:

Appendix yields that $\breve{\boldsymbol{p}}^{abc}$ contains the $(m+k)$th and the $(k-m)$th orders, both frequency orders contain the positive-, negative- and zero-sequence component, see Eq. (4.32). The elements in $\breve{\boldsymbol{p}}^{dq\pm(m+k)}$ and $\breve{\boldsymbol{p}}^{dq0(m+k)}$ are the same as in Eq. (4.30); $\breve{\boldsymbol{p}}^{dq\pm(k-m)}$and $\breve{\boldsymbol{p}}^{dq0(k-m)}$ can be seen in Eq. (4.33).

$$\begin{aligned} \breve{\boldsymbol{p}}^{abc} &= \breve{\boldsymbol{p}}^{abc+(m+k)} + \breve{\boldsymbol{p}}^{abc-(m+k)} + \breve{\boldsymbol{p}}^{abc0(m+k)} \\ &+ \breve{\boldsymbol{p}}^{abc+(k-m)} + \breve{\boldsymbol{p}}^{abc-(k-m)} + \breve{\boldsymbol{p}}^{abc0(k-m)} \\ &= \boldsymbol{T}^{2r/3s}_{+(m+k)}\breve{\boldsymbol{p}}^{dq+(m+k)} + \boldsymbol{T}^{2r/3s}_{-(m+k)}\breve{\boldsymbol{p}}^{dq-(m+k)} + \boldsymbol{T}^{2r/3s}_{0(m+k)}\breve{\boldsymbol{p}}^{dq0(m+k)} \\ &+ \boldsymbol{T}^{2r/3s}_{+(k-m)}\breve{\boldsymbol{p}}^{dq+(k-m)} + \boldsymbol{T}^{2r/3s}_{-(k-m)}\breve{\boldsymbol{p}}^{dq-(k-m)} + \boldsymbol{T}^{2r/3s}_{0(k-m)}\breve{\boldsymbol{p}}^{dq0(k-m)} \end{aligned} \tag{4.32}$$

where

$$\underbrace{\begin{bmatrix} \breve{\boldsymbol{p}}^{dq+(k-m)} \\ \breve{\boldsymbol{p}}^{dq-(k-m)} \\ \breve{\boldsymbol{p}}^{dq0(k-m)} \end{bmatrix}}_{\breve{\boldsymbol{p}}^{dq(k-m)}} = \underbrace{\begin{bmatrix} \frac{1}{2}\boldsymbol{F}_3(\boldsymbol{u}^{dq0m}) & \frac{1}{\sqrt{6}}\boldsymbol{F}_3(\boldsymbol{u}^{dq+m}) & \frac{1}{2}\boldsymbol{F}_3(\boldsymbol{u}^{dq-m}) \\ \frac{1}{\sqrt{6}}\boldsymbol{F}_3(\boldsymbol{u}^{dq-m}) & \frac{1}{2}\boldsymbol{F}_3(\boldsymbol{u}^{dq0m}) & \frac{1}{2}\boldsymbol{F}_3(\boldsymbol{u}^{dq+m}) \\ \frac{1}{3}\boldsymbol{F}_3(\boldsymbol{u}^{dq+m}) & \frac{1}{3}\boldsymbol{F}_3(\boldsymbol{u}^{dq-m}) & \frac{1}{2}\boldsymbol{F}_3(\boldsymbol{u}^{dq0m}) \end{bmatrix}}_{\boldsymbol{\mathcal{F}}_3(\boldsymbol{u}^{dqm})} \underbrace{\begin{bmatrix} \boldsymbol{i}^{dq+k} \\ \boldsymbol{i}^{dq-k} \\ \boldsymbol{i}^{dq0k} \end{bmatrix}}_{\boldsymbol{i}^{dqk}}$$

$$\text{with} \quad (m < k; 0 < m, k \leq M) \tag{4.33}$$

(3) Case $m = k$:

From the Appendix it can been observed that $\breve{\boldsymbol{p}}^{abc}$ contains the $(m+k)$th symmetrical components as well as a dc component $\breve{\boldsymbol{p}}^{abc}_{dc}$, where $\breve{\boldsymbol{p}}^{abc}_{dc} = [\breve{p}_{ab,dc} \quad \breve{p}_{bc,dc} \quad \breve{p}_{ca,dc}]^\top$.

$$\begin{aligned} \breve{\boldsymbol{p}}^{abc} &= \breve{\boldsymbol{p}}^{abc+(m+k)} + \breve{\boldsymbol{p}}^{abc-(m+k)} + \breve{\boldsymbol{p}}^{abc0(m+k)} + \breve{\boldsymbol{p}}^{abc}_{dc} \\ &= \boldsymbol{T}^{2r/3s}_{+(m+k)}\breve{\boldsymbol{p}}^{dq+(m+k)} + \boldsymbol{T}^{2r/3s}_{-(m+k)}\breve{\boldsymbol{p}}^{dq-(m+k)} + \boldsymbol{T}^{2r/3s}_{0(m+k)}\breve{\boldsymbol{p}}^{dq0(m+k)} + \breve{\boldsymbol{p}}^{abc}_{dc} \end{aligned} \tag{4.34}$$

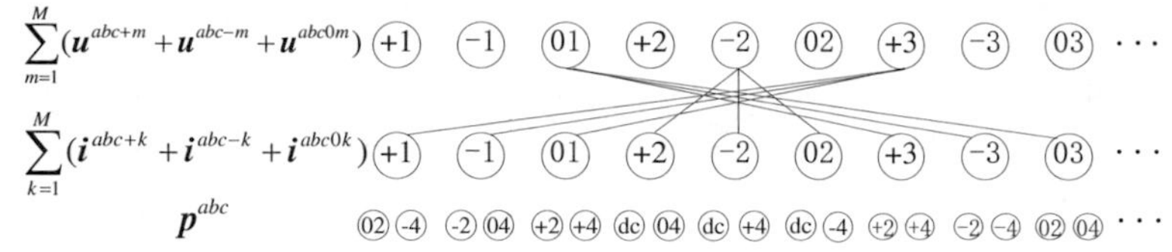

$\sum_{m=1}^{M}(\boldsymbol{u}^{abc+m}+\boldsymbol{u}^{abc-m}+\boldsymbol{u}^{abc0m})$ (+1) (−1) (01) (+2) (−2) (02) (+3) (−3) (03) ⋯

$\sum_{k=1}^{M}(\boldsymbol{i}^{abc+k}+\boldsymbol{i}^{abc-k}+\boldsymbol{i}^{abc0k})$ (+1) (−1) (01) (+2) (−2) (02) (+3) (−3) (03) ⋯

$\boldsymbol{p}^{abc}$ (02)(-4) (-2)(04) (+2)(+4) (dc)(04) (dc)(+4) (dc)(-4) (+2)(+4) (-2)(-4) (02)(04) ⋯

Figure 4.3.: Interaction example

where the elements in $\breve{\boldsymbol{p}}^{dq\pm(m+k)}$, $\breve{\boldsymbol{p}}^{dq0(m+k)}$ are the same as in Eq. (4.30); $\breve{\boldsymbol{p}}_{dc}^{abc}$, shown in Eq. (4.35), can also be expressed by the a function of $\boldsymbol{u}^{dqm}$ and $\boldsymbol{i}^{dqk}$.

$$\underbrace{\begin{bmatrix}\breve{p}_{ab,dc}\\ \breve{p}_{bc,dc}\\ \breve{p}_{ca,dc}\end{bmatrix}}_{\breve{\boldsymbol{p}}_{dc}^{abc}} = \underbrace{\begin{bmatrix}\sqrt{\frac{2}{3}} & 0 & \sqrt{\frac{2}{3}} & 0 & 1 & 0\\ -\frac{1}{\sqrt{6}} & \frac{1}{\sqrt{2}} & -\frac{1}{\sqrt{6}} & -\frac{1}{\sqrt{2}} & 1 & 0\\ -\frac{1}{\sqrt{6}} & \frac{1}{\sqrt{2}} & -\frac{1}{\sqrt{6}} & \frac{1}{\sqrt{2}} & 1 & 0\end{bmatrix}\boldsymbol{\mathcal{F}}_2(\boldsymbol{u}^{dqm})}_{\boldsymbol{\mathcal{F}}_0(\boldsymbol{u}^{dqm})}\boldsymbol{i}^{dqk}, \quad (m=k; 0<m,k\leq M) \tag{4.35}$$

The above cases show that one harmonic sequence component in the three-phase power is contributed by different combination of harmonic sequence interaction of the voltage and current, see Fig. 4.3. Take +4th in $\boldsymbol{p}^{abc}$ as an example, it can be generated by the interaction between $\boldsymbol{u}^{abc+3}$ and $\boldsymbol{i}^{abc01}$ (see $m=3$ and $k=1$ in Eq. (4.30)), or $\boldsymbol{u}^{abc-2}$ and $\boldsymbol{i}^{abc-2}$ (see $m=2$ and $k=2$ in Eqs. (4.30)), or $\boldsymbol{u}^{abc01}$ and $\boldsymbol{i}^{abc+3}$(see $m=1$ and $k=3$ in Eq. (4.30)), etc. Therefore one specific harmonic sequence in the power $\left(\boldsymbol{p}^{abc+n}, \boldsymbol{p}^{abc-n}, \boldsymbol{p}^{abc0n}, \forall n \in \{1,2,\cdots,M\}\right)$ is resulted from various combinations of harmonic sequence interaction of the branch voltage and current, the time-invariant representation for $\boldsymbol{p}^{abcn}$ would be as Eq. (4.36), the dc component in the power would be as Eq. (4.37).

$$\boldsymbol{p}^{dqn} = \boldsymbol{\mathcal{N}}(\boldsymbol{u}^{dq}, \boldsymbol{i}^{dq}) \tag{4.36}$$

$$\boldsymbol{p}_{dc}^{abc} = \boldsymbol{\mathcal{N}}_0(\boldsymbol{u}^{dq}, \boldsymbol{i}^{dq}) \tag{4.37}$$

where $\boldsymbol{\mathcal{N}}(\boldsymbol{u}^{dq}, \boldsymbol{i}^{dq})$ and $\boldsymbol{\mathcal{N}}_0(\boldsymbol{u}^{dq}, \boldsymbol{i}^{dq})$ are nonlinear terms expressed by the coupling of $\boldsymbol{u}^{dq}$ and $\boldsymbol{i}^{dq}$.

4.3.3. Observations of Branch Voltage Model

Replacing Eq. (4.14) and (4.17) into Eq. (4.10),

$$\boldsymbol{u}^{abc} = u_{C,dc}\sum_{k=1}^{M}(s^{abc+k} + s^{abc-k} + s^{abc0k}) \tag{4.38}$$

$$+\begin{bmatrix} \sum_{m=1}^{M}(u_{Cab+m} + u_{Cab-m} + u_{Cab0m})\sum_{k=1}^{M}(s_{ab+k} + s_{ab-k} + s_{ab0k}) \\ \sum_{m=1}^{M}(u_{Cbc+m} + u_{Cbc-m} + u_{Cbc0m})\sum_{k=1}^{M}(s_{bc+k} + s_{bc-k} + s_{bc0k}) \\ \sum_{m=1}^{M}(u_{Cca+m} + u_{Cca-m} + u_{Cca0m})\sum_{k=1}^{M}(s_{ca+k} + s_{ca-k} + s_{ca0k}) \end{bmatrix}$$

$$\boldsymbol{u}_C^{abcm} = \boldsymbol{u}_C^{abc+m} + \boldsymbol{u}_C^{abc-m} + \boldsymbol{u}_C^{abc0m}$$

$$= \boldsymbol{T}_{+m}^{2r/3s}\boldsymbol{u}_C^{dq+m} + \boldsymbol{T}_{-m}^{2r/3s}\boldsymbol{u}_C^{dq-m} + \boldsymbol{T}_{0m}^{2r/3s}\boldsymbol{u}_C^{dq0m} \tag{4.39}$$

$$\boldsymbol{s}^{abck} = \boldsymbol{s}^{abc+k} + \boldsymbol{s}^{abc-k} + \boldsymbol{s}^{abc0k} = \boldsymbol{T}_{+k}^{2r/3s}\boldsymbol{s}^{dq+k} + \boldsymbol{T}_{-k}^{2r/3s}\boldsymbol{s}^{dq-k} + \boldsymbol{T}_{0k}^{2r/3s}\boldsymbol{s}^{dq0k} \tag{4.40}$$

When looking into the second term of the right side of Eq. (4.38) and the right side of (4.25), they represent the coupling of two three-phase time-varying signals, both of these two signals contain multiple frequency components with sequences, thus the nth symmetrical components of Eq. (4.38) in dq frame would be

$$\boldsymbol{u}^{dqn} = u_{C,dc}\boldsymbol{s}^{dqn} + \boldsymbol{\mathcal{N}}(\boldsymbol{u}_C^{dq}, \boldsymbol{s}^{dq}) \tag{4.41}$$

where $\boldsymbol{\mathcal{N}}(\boldsymbol{u}_C^{dq}, \boldsymbol{s}^{dq})$ is a nonlinear term represented by the coupling of $\boldsymbol{u}_C^{dq}$ and $\boldsymbol{s}^{dq}$.

4.4. Derivation of Time-invariant Representation for Power Flow

The three-phase product signal $\breve{\boldsymbol{p}}^{abc}$ resulted from the coupling of $\boldsymbol{u}^{abcm}$ and $\boldsymbol{i}^{abck}$ in Eq. (4.26) contains $(m+k)$th and $|m-k|$th frequency orders when $m > k$ or $m < k$, $\breve{\boldsymbol{p}}^{abc}$ contains $(m+k)$th frequency order and dc components when $m = k$. In addition, the dc component and dq variables of the product can be expressed by functions of $\boldsymbol{u}^{dqm}$ and $\boldsymbol{i}^{dqk}$, which are the dq variables of $\boldsymbol{u}^{abcm}$ and $\boldsymbol{i}^{abck}$. Since $0 < m, k \leq M$, therefore $0 < m+k \leq 2M$ and $0 \leq |m-k| < M$, it can be derived that the highest frequency order of $\boldsymbol{p}^{abc}$ is $2M$, $\boldsymbol{p}^{abc}$ in Eq. (4.25) has the following generic description represented by the synchronous frame variables,

$$\begin{aligned}\boldsymbol{p}^{abc} &= \boldsymbol{p}_{dc}^{abc} + \sum_{n=1}^{2M} \boldsymbol{p}^{abcn} \\ &= \boldsymbol{p}_{dc}^{abc} + \sum_{n=1}^{2M} \left(\boldsymbol{T}_{+n}^{2r/3s} \boldsymbol{p}^{dq+n} + \boldsymbol{T}_{-n}^{2r/3s} \boldsymbol{p}^{dq-n} + \boldsymbol{T}_{0n}^{2r/3s} \boldsymbol{p}^{dq0n} \right)\end{aligned} \tag{4.42}$$

where the dc component $\boldsymbol{p}_{dc}^{abc}$ as well as the synchronous frame variables $\boldsymbol{p}^{dq+n}$, $\boldsymbol{p}^{dq-n}$, and $\boldsymbol{p}^{dq0n}$ contained in $\boldsymbol{p}^{dqn} = [\boldsymbol{p}^{dq+n\top} \quad \boldsymbol{p}^{dq-n\top} \quad \boldsymbol{p}^{dq0n\top}]^\top$ expressed by the coupling of $\boldsymbol{u}^{dq}$ and $\boldsymbol{i}^{dq}$ will be given in the next.

4.4.1. Time-invariant Representation for dc Component of Power Flow

From Section 4.3.2 it can be seen that only when $m = k$, it will generate dc component in the product, see Eq. (4.35). When all the frequencies $m = k = 1, 2, \cdots, M$ are considered, the time-invariant representation for the dc component in Eq. (4.42), i.e., the dc component of Eq. (4.25), is obtained

$$\boldsymbol{p}_{dc}^{abc} = \sum_{k=1}^{M} \boldsymbol{\mathcal{F}}_0(\boldsymbol{u}^{dqk})\boldsymbol{i}^{dqk} = \underbrace{\begin{bmatrix}\boldsymbol{\mathcal{F}}_0(\boldsymbol{u}^{dq1}) & \boldsymbol{\mathcal{F}}_0(\boldsymbol{u}^{dq2}) & \cdots & \boldsymbol{\mathcal{F}}_0(\boldsymbol{u}^{dqM})\end{bmatrix}}_{\boldsymbol{\mathcal{K}}(\boldsymbol{u}^{dq})} \underbrace{\begin{bmatrix}\boldsymbol{i}^{dq1} \\ \boldsymbol{i}^{dq2} \\ \vdots \\ \boldsymbol{i}^{dqM}\end{bmatrix}}_{\boldsymbol{i}^{dq}} \tag{4.43}$$

Until now, the term $\boldsymbol{\mathcal{N}}_0(\boldsymbol{u}^{dq}, \boldsymbol{i}^{dq})$ in Eq. (4.37) has been achieved.

4.4.2. Time-invariant Representation for Oscillating Component of Power Flow

It can be seen from Section 4.3.2, the three cases $m > k$, $m < k$ and $m = k$ can generate oscillating components in the product. All these three cases can contribute to generate a same frequency in the product signal in Eq. (4.25). When setting $m+k = n$ in Eq. (4.30) (see Eq. (4.44)), $m-k = n$ in (4.31) (see Eq. (4.45)) and $k-m=n$ in Eq. (4.33) (see Eq. (4.46)) while the prerequisites to derive Eqs. (4.30), (4.31), and (4.33) not being lost, it gives insight how these three cases contribute to generate the nth frequency component in the product signal.

$$\begin{aligned}
&\breve{\boldsymbol{p}}^{dq(m+k)} = \boldsymbol{\mathcal{F}}_1(\boldsymbol{u}^{dqm})\boldsymbol{i}^{dqk}, (m>k; m<k; m=k; 0<m, k\leq M)\\
&\xRightarrow{m+k=n} \breve{\boldsymbol{p}}^{dqn} = \boldsymbol{\mathcal{F}}_1(\boldsymbol{u}^{dq(n-k)})\boldsymbol{i}^{dqk},\ (n-k>k; n-k<k; n-k=k; 0<n-k, k\leq M)
\end{aligned} \tag{4.44}$$

$$\begin{aligned}
&\breve{\boldsymbol{p}}^{dq(m-k)} = \boldsymbol{\mathcal{F}}_2(\boldsymbol{u}^{dqm})\boldsymbol{i}^{dqk}, (m>k)\\
&\xRightarrow{m-k=n} \breve{\boldsymbol{p}}^{dqn} = \boldsymbol{\mathcal{F}}_2(\boldsymbol{u}^{dq(n+k)})\boldsymbol{i}^{dqk}, (n+k>k; 0<n+k, k\leq M)
\end{aligned} \tag{4.45}$$

$$\begin{aligned}
&\breve{\boldsymbol{p}}^{dq(k-m)} = \boldsymbol{\mathcal{F}}_3(\boldsymbol{u}^{dqm})\boldsymbol{i}^{dqk}, (m<k)\\
&\xRightarrow{k-m=n} \breve{\boldsymbol{p}}^{dqn} = \boldsymbol{\mathcal{F}}_3(\boldsymbol{u}^{dq(k-n)})\boldsymbol{i}^{dqk}, (k-n<k; 0<k-n, k\leq M)
\end{aligned} \tag{4.46}$$

Therefore $\boldsymbol{p}^{dqn}$ should be as the expression Eq. (4.47) when all feasible k in Eqs. (4.44) - (4.46) are taken into consideration. It has $1 \leq k \leq (n-1)$ in the first term of the right side of Eq. (4.47) because $n-k>0$ in Eq. (4.44); $1 \leq k \leq (M-n)$ in the second term because $n+k \leq M$ should be satisfied in Eq. (4.45) when the highest considered order is M; $(n+1) \leq k \leq M$ in the third term because $k-n>0$ should be satisfied in Eq. (4.46).

$$\boldsymbol{p}^{dqn} = \sum_{k=1}^{n-1} \boldsymbol{\mathcal{F}}_1(\boldsymbol{u}^{dq(n-k)})\boldsymbol{i}^{dqk} + \sum_{k=1}^{M-n} \boldsymbol{\mathcal{F}}_2(\boldsymbol{u}^{dq(n+k)})\boldsymbol{i}^{dqk} + \sum_{k=n+1}^{M} \boldsymbol{\mathcal{F}}_3(\boldsymbol{u}^{dq(k-n)})\boldsymbol{i}^{dqk} \tag{4.47}$$

In Eq. (4.47), $\boldsymbol{p}^{dqn}$ is the vector collecting the dq variables of the nth symmetrical components of Eq. (4.42), i.e., Eq. (4.25). Until now, the term $\boldsymbol{\mathcal{N}}(\boldsymbol{u}^{dq}, \boldsymbol{i}^{dq})$ in Eq. (4.36) has been obtained.

$$\underbrace{\begin{bmatrix} \boldsymbol{p}^{dq1} \\ \boldsymbol{p}^{dq2} \\ \vdots \\ \boldsymbol{p}^{dqM} \end{bmatrix}}_{\boldsymbol{p}^{dq}} = \underbrace{\begin{array}{c} \\ n=1 \\ n=2 \\ \vdots \\ n=M \end{array} \begin{array}{c} \begin{array}{cccc} k=1 & k=2 & \dots & k=M \end{array} \\ \begin{bmatrix} \boldsymbol{\mathcal{F}}_{1,1}(\boldsymbol{u}^{dq}) & \boldsymbol{\mathcal{F}}_{1,2}(\boldsymbol{u}^{dq}) & \dots & \boldsymbol{\mathcal{F}}_{1,M}(\boldsymbol{u}^{dq}) \\ \boldsymbol{\mathcal{F}}_{2,1}(\boldsymbol{u}^{dq}) & \boldsymbol{\mathcal{F}}_{2,2}(\boldsymbol{u}^{dq}) & \dots & \boldsymbol{\mathcal{F}}_{2,M}(\boldsymbol{u}^{dq}) \\ \vdots & \vdots & \ddots & \vdots \\ \boldsymbol{\mathcal{F}}_{M,1}(\boldsymbol{u}^{dq}) & \boldsymbol{\mathcal{F}}_{M,2}(\boldsymbol{u}^{dq}) & \dots & \boldsymbol{\mathcal{F}}_{M,M}(\boldsymbol{u}^{dq}) \end{bmatrix} \end{array}}_{\boldsymbol{\mathcal{F}}(\boldsymbol{u}^{dq})} \underbrace{\begin{bmatrix} \boldsymbol{i}^{dq1} \\ \boldsymbol{i}^{dq2} \\ \vdots \\ \boldsymbol{i}^{dqM} \end{bmatrix}}_{\boldsymbol{i}^{dq}} \tag{4.48}$$

The above compact expression as Eq. (4.48) can be obtained, when putting the dq variables for symmetrical components of all the considered frequencies in $\boldsymbol{p}^{abc}$ into a vector. $\boldsymbol{\mathcal{F}}(\boldsymbol{u}^{dq})$ consists of $M \times M$ submatrix blocks, each block is with dimension 6×6 and denoted by $\boldsymbol{\mathcal{F}}_{n,k}(\boldsymbol{u}^{dq})$ where n and k represent the block position in the matrix $\boldsymbol{\mathcal{F}}(\boldsymbol{u}^{dq})$, the submatrix is as

$$\boldsymbol{\mathcal{F}}_{n,k}(\boldsymbol{u}^{dq}) = \begin{cases} \boldsymbol{\mathcal{F}}_1(\boldsymbol{u}^{dq(n-k)}) + \boldsymbol{\mathcal{F}}_2(\boldsymbol{u}^{dq(n+k)}), & (n > k) \quad (4.49a) \\ \boldsymbol{\mathcal{F}}_2(\boldsymbol{u}^{dq(n+k)}), & (n = k) \quad (4.49b) \\ \boldsymbol{\mathcal{F}}_3(\boldsymbol{u}^{dq(k-n)}) + \boldsymbol{\mathcal{F}}_2(\boldsymbol{u}^{dq(n+k)}), & (n < k) \quad (4.49c) \end{cases}$$

It should be clear that in Eqs. (4.49a)-(4.49c) if the harmonic order $(n + k)$ in $\boldsymbol{\mathcal{F}}_2(\boldsymbol{u}^{dq(n+k)})$ is larger than the highest considered order M, such harmonic components can be set to $\mathbf{0}$.

4.5. Utilisation of Harmonic Interaction Analysis

4.5.1. Time-invariant Model Based on Harmonic Interaction Analysis

As mentioned in Section 4.3, the derivation procedure for $\boldsymbol{\mathcal{N}}(\boldsymbol{u}^{dq}, \boldsymbol{i}^{dq})$ in Eq. (4.36) can be applied to used to derive $\boldsymbol{\mathcal{N}}(\boldsymbol{u}_C^{dq}, \boldsymbol{s}^{dq})$ in Eq. (4.41). $\boldsymbol{\mathcal{N}}(\boldsymbol{u}^{dq}, \boldsymbol{i}^{dq})$ in Eq. (4.36) has been derived in Section 4.4 (Eq. (4.47)), therefore $\boldsymbol{\mathcal{N}}(\boldsymbol{u}_C^{dq}, \boldsymbol{s}^{dq})$ can be obtained by application of the similar derivation procedure. Eq. (4.41)becomes

$$\begin{aligned} \boldsymbol{u}^{dqn} &= u_{C,dc}\boldsymbol{s}^{dqn} + \boldsymbol{\mathcal{N}}(\boldsymbol{u}_C^{dq}, \boldsymbol{s}^{dq}) = u_{C,dc}\boldsymbol{s}^{dqn} \\ &+ \sum_{k=1}^{n-1} \boldsymbol{\mathcal{F}}_1(\boldsymbol{u}_C^{dq(n-k)})\boldsymbol{s}^{dqk} + \sum_{k=1}^{M-n} \boldsymbol{\mathcal{F}}_2(\boldsymbol{u}_C^{dq(n+k)})\boldsymbol{s}^{dqk} + \sum_{k=n+1}^{M} \boldsymbol{\mathcal{F}}_3(\boldsymbol{u}_C^{dq(k-n)})\boldsymbol{s}^{dqk} \end{aligned} \quad (4.50)$$

With the application of the compact description as Eq. (4.48), the time-invariant representation of the osicllating component in Eq. (4.38) can be achieved as Eq. (4.51). The reason why it is unnecessary to include an expression for a dc component of Eq. (4.38) will be explained in Section 4.6.

$$\boldsymbol{u}^{dq} = u_{C,dc}\boldsymbol{s}^{dq} + \boldsymbol{\mathcal{F}}(\boldsymbol{u}_C^{dq})\boldsymbol{s}^{dq} \quad (4.51)$$

Eqs. (4.52a)-(4.52c) define a time-invariant state-space representation of the average model of Eqs. (4.7) and (4.9), with application of the derived expression in Eq. (4.51), (4.43), and (4.48).

$$\frac{d}{dt}\boldsymbol{i}^{dq} = \left(\boldsymbol{J} - \frac{r}{L}\boldsymbol{I}_{6M}\right)\boldsymbol{i}^{dq} - \frac{1}{L}\underbrace{\left(u_{C,dc}\boldsymbol{I}_{6M} + \boldsymbol{\mathcal{F}}(\boldsymbol{u}_C^{dq})\right)\boldsymbol{s}^{dq}}_{\boldsymbol{u}^{dq}} + \frac{1}{L}\boldsymbol{v}_s^{dq} \tag{4.52a}$$

$$\frac{d}{dt}\boldsymbol{w}_{dc}^{abc} = \underbrace{\boldsymbol{\mathcal{K}}(\boldsymbol{u}^{dq})\boldsymbol{i}^{dq}}_{\boldsymbol{p}_{dc}^{abc}} \tag{4.52b}$$

$$\frac{d}{dt}\boldsymbol{w}^{dq} = \boldsymbol{J}\boldsymbol{w}^{dq} + \underbrace{\boldsymbol{\mathcal{F}}(\boldsymbol{u}^{dq})\boldsymbol{i}^{dq}}_{\boldsymbol{p}^{dq}} \tag{4.52c}$$

where

$$\boldsymbol{J} = \text{blkdiag}(\boldsymbol{J}_{\omega}, \boldsymbol{J}_{2\omega}, \cdots, \boldsymbol{J}_{M\omega}) \tag{4.53}$$

$$\boldsymbol{J}_{n\omega} = \text{blkdiag}\left(\begin{bmatrix} 0 & 1 \\ -1 & 0 \end{bmatrix}, \begin{bmatrix} 0 & 1 \\ -1 & 0 \end{bmatrix}, \begin{bmatrix} 0 & 1 \\ -1 & 0 \end{bmatrix}\right) n\omega, \quad (n = 1, 2, \cdots, M) \tag{4.54}$$

Eqs. (4.52a)-(4.52c) preserve the dynamics and harmonic interaction in the model Eqs. (4.7)-(4.9), and inherit the same limitations as the analytical average models in the stationary frame. As for any other analytical average model, this implies that the developed time-invariant representation is not representing any physical saturation limits within the model, like for instance the overmodulation limit that can be reached if the voltage reference for the converter is higher than the available voltage in the internal capacitors. As long as the delta-connected CHB is operated within its limitations, such as the overmodulation limits, the derived model will contain detailed information about the operation of the converter, including the harmonic coupling between the sequences of various frequency components.

4.5.2. Determination of Control Variables

In order to achieve the quantification of the variables at steady state, the reference branch currents should be determined first, once the reference branch currents are determined, the reference branch voltages can be calculated. With the reference branch voltage and currents, the reference power flows can be predicted. With this predicted power flows, the reference capacitor voltage harmonics can be calculated based on Eq. (4.18). Once the capacitor voltage harmonics can be calculated, the accurate switching function calculation can be achieved.

Reference branch currents and branch voltages determination

The following relations exist among the branch currents, the source currents, the loads currents and the circulating current, where $*$ indicates the reference value:

$$\begin{cases} i_{ab}^* = \frac{(i_{sa}^* - i_{La}) - (i_{sb}^* - i_{Lb})}{3} + i_0^* \\ i_{bc}^* = \frac{(i_{sb}^* - i_{Lb}) - (i_{sc}^* - i_{Lc})}{3} + i_0^* \\ i_{ca}^* = \frac{(i_{sc}^* - i_{Lc}) - (i_{sa}^* - i_{La})}{3} + i_0^* \end{cases} \tag{4.55}$$

where the desired source currents $i_{sa,sb,sc}^*$ can be calculated from source current strategies, such as the UPF, PHC, or other strategies as mentioned in Chapter 3. The load currents are measurable, thus the only unknown variable for reference branch current determination is the reference circulating current i_0^*, in other words, the only unknown variables in $\boldsymbol{i}^{*dq}$, also the reference of the state variable of Eq. (4.52a), are dq values of the circulating current (zero sequence current). As mentioned in Section 4.2, the dc component in the powers as in Eq. (4.10) should be 0 at steady state to avoid drifting the capacitor energy. It means that the reference branch currents and voltages should make Eq. (4.52b) equal to $\mathbf{0}$,

$$\boldsymbol{p}_{dc}^{*abc} = \begin{bmatrix} p_{ab,dc}^* & p_{bc,dc}^* & p_{ca,dc}^* \end{bmatrix}^\top = \boldsymbol{\mathcal{K}}(\boldsymbol{u}^{*dq})\boldsymbol{i}^{*dq} = \begin{bmatrix} 0 & 0 & 0 \end{bmatrix}^\top \tag{4.56}$$

Since $\boldsymbol{u}^{*dq}$ cannot be calculated if $\boldsymbol{i}^{*dq}$ is unknown, considering that the branch resistance and inductance are very small, $\boldsymbol{p}_{dc}^{*abc} = \boldsymbol{\mathcal{K}}(\boldsymbol{u}^{*dq})\boldsymbol{i}^{*dq} = \mathbf{0}$ can be approximated as

$$\boldsymbol{p}_{dc}^{*abc} \approx \boldsymbol{\mathcal{K}}(\boldsymbol{v}_s^{dq})\boldsymbol{i}^{*dq} = \mathbf{0} \tag{4.57}$$

to calculate the dq variables of the circulating current so that the dq variables of reference branch currents can be determined. By setting $\frac{d}{dt}\boldsymbol{i}^{*dq} = \mathbf{0}$ (Eq. (4.52a)), $\boldsymbol{u}^{*dq}$ can be calculated by

$$\boldsymbol{u}^{*dq} = L\left(\boldsymbol{J} - \frac{r}{L}\boldsymbol{I}_{6M}\right)\boldsymbol{i}^{*dq} + \boldsymbol{v}_s^{dq} \tag{4.58}$$

Reference Power Flows

As mentioned before, Eq. (4.48) is the time-invariant representation of the oscillating components of Eq. (4.25), the following gives the coupling effect presented in Eq. (4.10) with reference values,

$$\boldsymbol{p}^{*dq} = \boldsymbol{\mathcal{F}}(\boldsymbol{u}^{*dq})\boldsymbol{i}^{*dq} \tag{4.59}$$

This has not been done before in other literatures while other literatures often only give the average and/or 2nd order power component when the branch voltage and current contain only the fundamental frequency such as in STATCOM application, i.e., other literatures only give the following when the symbols in this disseratation are used

$$\text{average power} \quad \boldsymbol{p}_{dc}^{*abc} = \boldsymbol{\mathcal{F}}_0\left(\boldsymbol{u}^{*dq1}\right)\boldsymbol{i}^{*dq1} \tag{4.60}$$

$$\text{2nd order power} \quad \boldsymbol{p}^{*dq2} = \boldsymbol{\mathcal{F}}_1\left(\boldsymbol{u}^{*dq1}\right)\boldsymbol{i}^{*dq1} \tag{4.61}$$

Relation between Power Flows and SM Capacitor Voltages

The following can be obtained by putting Eq. (4.14) with the reference values to Eq. (4.18),

$$\begin{aligned}
\begin{bmatrix} p_{ab}^* \\ p_{bc}^* \\ p_{ca}^* \end{bmatrix} &= C_{sum} \begin{bmatrix} u_{Cab}^* \cdot \frac{d}{dt}u_{Cab}^* \\ u_{Cbc}^* \cdot \frac{d}{dt}u_{Cbc}^* \\ u_{Cca}^* \cdot \frac{d}{dt}u_{Cca}^* \end{bmatrix} \\
&= C_{sum} \left\{ u_{C,dc}^* \begin{bmatrix} \sum_{n=1}^{M} \frac{d}{dt}u_{Cabn}^* \\ \sum_{n=1}^{M} \frac{d}{dt}u_{Cbcn}^* \\ \sum_{n=1}^{M} \frac{d}{dt}u_{Ccan}^* \end{bmatrix} + \begin{bmatrix} \left(\sum_{n=1}^{M} u_{Cabn}^*\right) \cdot \left(\sum_{n=1}^{M} \frac{d}{dt}u_{Cabn}^*\right) \\ \left(\sum_{n=1}^{M} u_{Cbcn}^*\right) \cdot \left(\sum_{n=1}^{M} \frac{d}{dt}u_{Cbcn}^*\right) \\ \left(\sum_{n=1}^{M} u_{Ccan}^*\right) \cdot \left(\sum_{n=1}^{M} \frac{d}{dt}u_{Ccan}^*\right) \end{bmatrix} \right\}
\end{aligned} \tag{4.62}$$

Compare the second term of the right side of Eq. (4.62) with (4.25), it can be found that both represents the product of two three-phase signals, each with multiple frequencies with symmetrical components. Therefore the derivation procedure of Eq. (4.48) from Eq. (4.25) can also be applied to obtain a time-invariant representation of the second term of the right side of Eq. (4.62), resulting in

$$\boldsymbol{p}^{*dq} = C_{sum}\left\{u_{C,dc}^* \boldsymbol{u}_C^{\prime *dq} + \boldsymbol{\mathcal{F}}\left(\boldsymbol{u}_C^{*dq}\right)\boldsymbol{u}_C^{\prime *dq}\right\} \tag{4.63}$$

where vectors $\boldsymbol{u}_C^{\prime *dq}$ and $\boldsymbol{u}_C^{*dq}$ consists of dq values of $\frac{d}{dt}\boldsymbol{u}_C^{*abc}$ and $\boldsymbol{u}_C^{*abc}$ respectively, defined by

$$\boldsymbol{u}_C^{\prime *dq} = [\boldsymbol{u}_C^{\prime *dq1\top} \quad \boldsymbol{u}_C^{\prime *dq2\top} \quad \cdots \quad \boldsymbol{u}_C^{\prime *dqM\top}]^\top \tag{4.64}$$

$$\boldsymbol{u}_C^{*dq} = [\boldsymbol{u}_C^{*dq1\top} \quad \boldsymbol{u}_C^{*dq2\top} \quad \cdots \quad \boldsymbol{u}_C^{*dqM\top}]^\top \tag{4.65}$$

There is

$$\frac{d}{dt}\boldsymbol{u}_C^{*abc+n} = \boldsymbol{T}_{+n}^{2r/3s}\boldsymbol{u}_C^{\prime *dq+n}, \quad \frac{d}{dt}\boldsymbol{u}_C^{*abc-n} = \boldsymbol{T}_{-n}^{2r/3s}\boldsymbol{u}_C^{\prime *dq-n}, \quad \frac{d}{dt}\boldsymbol{u}_C^{*abc0n} = \boldsymbol{T}_{0n}^{2r/3s}\boldsymbol{u}_C^{\prime *dq0n}$$

$$\boldsymbol{u}_C^{*abc+n} = \boldsymbol{T}_{+n}^{2r/3s}\boldsymbol{u}_C^{*dq+n}, \quad \boldsymbol{u}_C^{*abc-n} = \boldsymbol{T}_{-n}^{2r/3s}\boldsymbol{u}_C^{*dq-n}, \quad \boldsymbol{u}_C^{*abc0n} = \boldsymbol{T}_{0n}^{2r/3s}\boldsymbol{u}_C^{*dq0n}$$

The relationship between $\boldsymbol{u}_C^{*dq}$ and $\boldsymbol{u}_C^{\prime *dq}$ is then,

$$\boldsymbol{u}_C^{\prime *dq} = -\boldsymbol{J}\boldsymbol{u}_C^{*dq} \tag{4.66}$$

where $\boldsymbol{J}$ is as Eq. (4.53), thus Eq. (4.63) becomes

$$\boldsymbol{p}^{*dq} = -C_{sum}\left\{u_{C,dc}^{*}\boldsymbol{I}_{6M} + \boldsymbol{\mathcal{F}}(\boldsymbol{u}_C^{*dq})\right\}\boldsymbol{J}\boldsymbol{u}_C^{*dq} \tag{4.67}$$

Putting the calculated powers from Eq. (4.59) to Eq. (4.67), then Eq. (4.67) becomes a nonlinear equation with unknown variables $\boldsymbol{u}_C^{*dq}$. The time-invariant expression Eq. (4.67) represents an accurate relation between the power flows and SM capacitor voltages. Solving the nonlinear equation Eq. (4.67) can result in a precise calculation of $\boldsymbol{u}_C^{*dq}$, which is more accurate than conventional capacitor voltage calculation [WCT+16], where the instantaneous power is approximated as $p_{ab,bc,ca}^{*} = C_{sum}u_{C,dc}^{*}\frac{d}{dt}u_{Cab,Cbc,Cca}^{*}$.

Reference Switching Functions in Harmonic Sequences

Once $\boldsymbol{u}_C^{*dq}$ is obtained, the reference switching functions in dq-frame can then be calculated based on Eq. (4.52a),

$$\boldsymbol{s}^{*dq} = \left\{u_{C,dc}^{*}\boldsymbol{I}_{6M} + \boldsymbol{\mathcal{F}}\left(\boldsymbol{u}_C^{*dq}\right)\right\}^{-1}\boldsymbol{u}^{*dq} \tag{4.68}$$

The calculation of switching functions from Eq. (4.68) is more accurate than the conventional calculation [LWY+17, KV17], where the reference branch output voltage is approximated as $u_{ij}^{*} = u_{C,dc}^{*}s_{ij}^{*}$.

4.6. Validation via Time-Domain Simulation

The delta-connected CHB multilevel converter simulation model has been built in MATLAB/Simulink. The simulation parameters are listed in Table 4.1. It should be noticed that the individual SM capacitance in Table 4.1 is much smaller than that in Chapter 3, see Table 3.2. The capacitor voltage ripple in this chapter is about ±25%. The reference SM capacitor sum voltage is $u^{*}_{C,dc} = 62kV$. Fig. 4.4 shows the diagram of the whole system for validation. It should be clear that the PSPWM strategy, the reference source current strategy, and the LQR controller are only for validation, other PWM strategies, source current strategies and control schemes can also be applied.

The simulation results of the detailed harmonics in the power, capacitor voltage

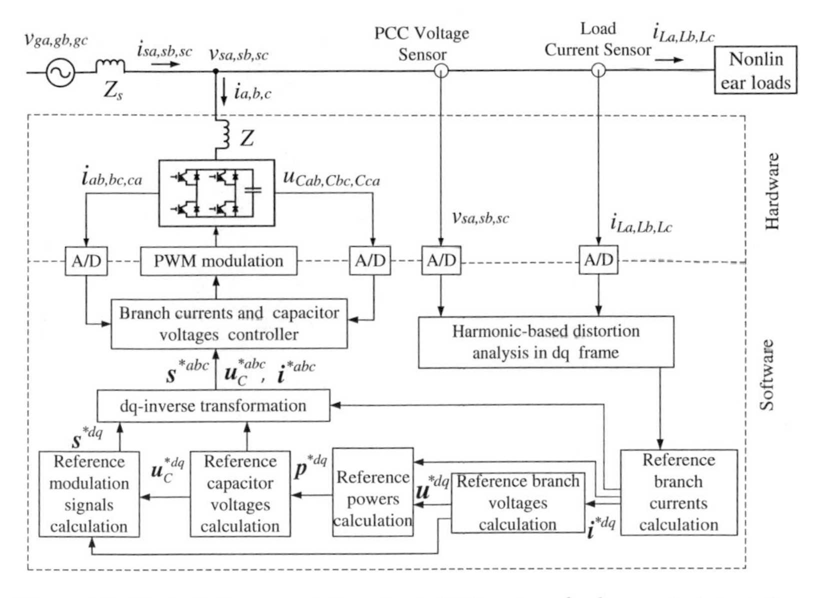

Figure 4.4.: Control diagram of the shunt APF system for harmonic interaction analysis validation

Table 4.1.: Parameters used in simulations

Parameter	Symbol	Value
Rated line frequency	$\omega/2\pi$	50 Hz
DC capacitor of full-bridge submodule	C	50 μF
Carrier frequency for PSPWM	f_c	5000 Hz
Branch inductor	L	9.2 mH
Branch resistor	r	0.0566 Ω
Number of SMs in one branch	N	8
APF nominal rating	S_N	25 MVA

and switching function are compared with the calculation results of the proposed approach, indicating the high accuracy. The proposed calculation of capacitor voltage by using Eq. (4.67) can be achieved via MATLAB 'fsolve' or 'lsqnonlin'. Moreover, to show the superiority of the proposed approach, the comparison of capacitor voltage and switching function calculation in the proposed method with the conventional method has been made- the conventional calculation of the capacitor voltage ripple without considering the harmonic interaction between the capacitor voltage and its derivation [WCT+16, NL07]. Using the symbols in this chapter, the conventional calculation of the capacitor voltage in [WCT+16, NL07] is expressed by

$$\boldsymbol{u}_C^{*dq} = -\frac{1}{C_{sum} u_{C,dc}^*} \boldsymbol{J}^{-1} \boldsymbol{p}^{*dq} \tag{4.69}$$

The comparison of the switching function calculation from the proposed and conventional method [LWY+17, KV17] is also presented. The expression of the conventional calculation of the switching function using the symbols in this chapter can be expressed by

$$\boldsymbol{s}^{*dq} = \frac{1}{u_{C,dc}^*} \boldsymbol{u}^{*dq} \tag{4.70}$$

It can be seen from Eq. (4.69) that the conventional calculation of the capacitor voltage is only capable to present harmonic orders in the branch power, while Eq. (4.70) shows that conventional calculation of the switching signal is unable to present harmonic orders not in the branch voltage.

The author in [Wak10] mentions that odd harmonics are the characteristic harmonic components in today's power networks, due to that a property known as half-wave symmetry exists in the present infrastructures, $f(t) = -\left(t \pm \frac{T}{2}\right)$, which leads to zero dc components and cancellation of even order harmonics, therefore in the load current $i_{La,Lb,Lc}$ and the PCC voltage $v_{sa,sb,sc}$ there is only odd order line-frequency components. This will result in the harmonics in some variables, for instance the power flows and the SM capacitor voltages, only even orders while in other variables, like switching functions and branch currents/voltages, only odd orders. The coupling of capacitor voltages with even-order harmonics and switching functions with odd-order result in that no matter under transient or steady state, the branch currents/voltages contain only odd order frequencies without dc component. Thus it is unnecessary to express dc component for branch currents/voltages in Section 4.5. It should be noticed that the proposed harmonic interaction analysis can still be applied as long as setting the nonexisting harmonics to 0.

Five cases are presented here,

Case A The PCC voltages are ideal as Eq. (3.94) while the load currents contain only the fundamental frequency component as Eq. (4.71)

Case B The PCC voltages are ideal as Eq. (3.94) while the load currents contain fundamental frequency and harmonic components as Eq. (3.81)

Case C The PCC voltages are unbalanced and undistorted as Eq. (3.95) while the load currents contain fundamental frequency and harmonic components as Eq. (3.81)

Case D The PCC voltages are balanced and distorted as Eq. (3.96) while the load currents contain fundamental frequency and harmonic components as Eq. (3.81)

Case E The PCC voltages are unbalanced and distorted as Eq. (3.97) while the load currents contain fundamental frequency and harmonic components as Eq. (3.81)

4.6.1. Case A

The PCC voltages in kV are as Eq. (3.94) and the load currents in kA are as Eq. (4.71), containing only the positive-sequence fundamental-frequency component,

$$\begin{cases} i_{La} = \sin\left(\omega t - \frac{\pi}{18}\right) \\ i_{Lb} = \sin\left(\omega t - \frac{13\pi}{18}\right) \\ i_{Lc} = \sin\left(\omega t + \frac{11\pi}{18}\right) \end{cases} \tag{4.71}$$

The line-line voltages in this case satisfy

$$\hbar_{v_{sij}} \in \{+1\} \tag{4.72}$$

where $\hbar_{(\bullet)}$ represents a set collecting the harmonic sequences of the variable in the subscript. For reactive power compensation, the steady branch currents as well as the steady branch voltages have the following harmonic sequences

$$\hbar_{i^*_{ij}}, \hbar_{u^*_{ij}} \in \{+1\} \tag{4.73}$$

The steady circulating current is 0. Fig. 4.5. (a) shows the comparison of the simulation results and the proposed calculation of the amplitude of each frequency order power components in positive, negative and zero sequence, denoted respectively by P_{+n}, P_{-n}, P_{0n}, calculated from $P_{\pm n} = \sqrt{\frac{2}{3}(p^2_{d\pm n} + p^2_{q\pm n})}$ and $P_{0n} = \sqrt{p^2_{d0n} + p^2_{q0n}}$ and plotted in per unit quantities, with base values as the converter nominal rating in Table 4.1. Clearly, the proposed power calculation (Eq. (4.59)) shows

$$\hbar_{p^*_{ij}} \in \{-2\} \tag{4.74}$$

Table 4.2 shows the comparison results of the amplitude of the sum of SM capacitor voltages in harmonic positive, negative and zero sequence, denoted respectively by U_{C+n}, U_{C-n}, and U_{C0n}, calculated from $U_{C\pm n} = \sqrt{\frac{2}{3}(u^2_{Cd\pm n} + u^2_{Cq\pm n})}$ and $U_{C0n} = \sqrt{u^2_{Cd0n} + u^2_{Cq0n}}$. The errors in Table 4.2 are calculated by the deviation between the simulation results and the proposed/conventional calculations devided by the simulation results. It can be seen that the errors of the proposed approach are lower than the conventional calculation, evidently indicating the accuracy of the proposed method. The conventional calculation (Eq. (4.69)) shows $\hbar_{u^*_{Cij}} = \hbar_{p^*_{ij}} \in$

$\{-2\}$, since the conventional calculation of SM capacitor voltages is unable to present frequency orders not in the power flows, while the proposed capacitor voltage calculation (Eq. (4.67)) indicates that

$$\hbar_{u^*_{Cij}} \in \{-2, +4, 06, -8, \cdots\} \tag{4.75}$$

where −2th, +4th, and 06th are dominate and others are negligible, which is very close to the simulation results.

Fig. 4.5. (b) gives the comparison results of the amplitude of the sequences in each frequency order switching signal from simulation, the proposed and conventional calculation. In Fig. 4.5. (b) 'simulation (without PSPWM)' represents the reference each submodule uses to compare with carriers, while the 'simulation (with PSPWM)' represents the stepped multilevel waveforms from modulation, plotted in per unit with base value as N, since the stepped waveforms belong to $\{-N, -N+1, \cdots, N-1, N\}$. The amplitude of each frequency order switching functions in positive, negative and zero sequence, denoted respectively by S_{+n}, S_{-n}, S_{0n}, are calculated from $S_{\pm n} = \sqrt{\frac{2}{3}(s^2_{d\pm n} + s^2_{q\pm n})}$ and $S_{0n} = \sqrt{s^2_{d0n} + s^2_{q0n}}$. In the proposed method the switching function harmonics until 19th are presented. The proposed calculation (Eq. (4.68)) shows that

$$\hbar_{s^*_{ij}} \in \{+1, 03, -5, +7, \cdots\} \tag{4.76}$$

while conventionally $\hbar_{s^*_{ij}} = \hbar_{u^*_{ij}} \in \{+1\}$ since the conventional calculation of switching functions (Eq. (4.70)) is unable to present frequency orders not in the branch voltages. It is obvious that the error of the proposed calculation is lower, indicating that it is closer to the simulation results. Comparing with +1st, 03rd, −5th, the values of other harmonics in the switching function are already very small, therefore, the errors of such harmonics are relatively larger, but it is still acceptable.

Fig. 4.6 shows the waveforms of the sum of SM capacitor voltages in each branch and switching functions/modulation signals in time-domain. It is obvious to see that the waveforms from the proposed method are closer to the simulated results, proving its high accuracy, while the waveforms of the conventional waveforms show distinct errors.

4.6.2. Case B

The PCC voltages in Case B are as Eq. (3.94) in kV, and the load currents are as Eq. (3.81) in kA. In this case the steady circulating current is 0, see Section 3.7.1. In order to compensate the load current harmonics and reactive power the steady branch currents and steady branch voltages have following characteristics

$$\hbar_{i^*_{ij}}, \hbar_{u^*_{ij}} \in \{+1, -5, +7\} \tag{4.77}$$

Fig. 4.7. (a) shows the comparison of power components in per unit, which shows

$$\hbar_{p^*_{ij}} \in \{-2, +4, 06, -8, +10, 012, -14\} \tag{4.78}$$

and the proposed calculation is close to the simulation. The submodule capacitor voltage values in the proposed calculation belong to

$$\hbar_{u^*_{Cij}} \in \{-2, +4, 06, -8, \cdots\} \tag{4.79}$$

see Table 4.3, which gives the comparison results of the amplitude of the capacitor voltages in harmonic symmetrical components until 16th (only several volts higher than 16th). The listed results show that the proposed calculation is able to solve more existing harmonics and more accurate than the conventional calculation.

Fig. 4.7. (b) shows that the proposed calculation of the switching function

$$\hbar_{s^*_{ij}} \in \{+1, 03, -5, +7, 09, \cdots\} \tag{4.80}$$

while in the conventional calculation $\hbar_{s^*_{ij}} = \hbar_{u^*_{ij}} \in \{+1, -5, +7\}$. It is obvious that the proposed calculation is closer to the simulation, proving its superiority than the conventional method.

The waveforms of capacitor voltages and switching signals shown in Fig. 4.8 also illustrate the accuracy and superiority of the proposed calculation.

4.6.3. Case C

The harmonic free and unbalanced PCC voltages are as Eq. (3.95) in kV and the load currents are as Eq. (3.81) in kA. Harmonic sequences in PCC line-line voltages are

$$\hbar_{v_{sij}} \in \{\pm 1\} \tag{4.81}$$

The circulating current is non-zero but of fundamental frequency, see Section 3.7.2. To achieve the PHC source currents, the steady branch currents and voltages have characteristics as

$$\hbar_{i^*_{ij}} \in \{+1, 01, -5, +7\}, \quad \hbar_{u^*_{ij}} \in \{\pm 1, 01, -5, +7\} \tag{4.82}$$

From Fig. 4.9.(a) it can be seen that the power components in the proposed calculation

$$\hbar_{p^*_{ij}} \in \{\pm 2, 02, \pm 4, 04, \pm 6, 06, \pm 8, 08, +10, 012, -14\} \tag{4.83}$$

which is very close to the simulated results. Table 4.4 shows that the proposed calculation of the capacitor voltages results in

$$\hbar_{u^*_{Cij}} \in \{\pm 2, 02, \pm 4, 04, \pm 6, 06, \cdots\} \tag{4.84}$$

while conventionally $\hbar_{u^*_{Cij}} = \hbar_{p^*_{ij}}$. Table 4.4 proves the accuracy of the proposed calculation. The harmonics higher than 16th in capacitor voltages are very small. The harmonic sequences existed in the proposed capacitor voltages like -10th, 010th and ± 12th but not in the conventional SM capacitor voltage calculation prove that the proposed calculation can solve more existing harmonics.

Fig. 4.9.(b) shows the proposed switching signals

$$\hbar_{s^*_{ij}} \in \{\pm 1, 01, \pm 3, 03, \pm 5, 05, \cdots\} \tag{4.85}$$

proving its accuracy, while conventionally $\hbar_{s^*_{ij}} = \hbar_{u^*_{ij}}$.

Fig. 4.10 gives the time-domain capacitor voltages and switching functions, validating that compared with the conventional calculations, the proposed calculations are closer to the simulation results, since in the conventional method the harmonic interaction is neglected.

4.6.4. Case D

In this case, the PCC voltages are balanced but contain harmonics (Eq. (3.96) in kV) and the load currents are as Eq. (3.81) in kA. The line-line voltages

$$\hbar_{v_{sij}} \in \{+1, -5, +7\} \tag{4.86}$$

The reference circulating current is 0, see Section 3.7.3. The harmonic sequence sets of the branch currents and the branch voltages are

$$\hbar_{i^*_{ij}}, \hbar_{u^*_{ij}} \in \{+1, -5, +7\} \tag{4.87}$$

At steady state the branch powers have the following harmonic sequences

$$\hbar_{p^*_{ij}} \in \{-2, +4, 06, -8, +10, 012, -14\} \tag{4.88}$$

close to the simulate results, see Fig. 4.11.(a). The harmonic sequence set of the reference capacitor voltages from the proposed calculation is shown as

$$\hbar_{u^*_{Cij}} \in \{-2, +4, 06, -8, +10, 012, \ldots\} \tag{4.89}$$

which is validated by Table 4.5. Compared with the conventional calculation, the proposed calculation is closer to the simulation.

The switching functions in the proposed calculation satisfy

$$\hbar_{s^*_{ij}} \in \{+1, 03, -5, +7, 09, \cdots\} \tag{4.90}$$

see Fig. 4.11.(b), while in the conventional calculation $\hbar_{s^*_{ij}} = \hbar_{u^*_{ij}} \in \{+1, -5, +7\}$. The proposed calculation is more accurate than the conventional calculation. The time-domain waveforms of capacitor voltages and switching functions shown in Fig. 4.12 indicate the accuracy of the proposed calculation.

4.6.5. Case E

In this case, the PCC voltages as shown in Eq. (3.97) in kV are unbalanced and distorted. The harmonic sequences of the PCC line-line voltages are

$$\hbar_{v_{sij}} \in \{\pm 1, \pm 5, \pm 7\} \tag{4.91}$$

The circulating current contains the fundamental frequency component, see Section 3.7.4. The harmonic sets of the reference branch currents as well as branch voltages can be seen in

$$\hbar_{i^*_{ij}} \in \{+1, 01, -5, +7\}\,, \quad \hbar_{u^*_{ij}} \in \{\pm 1, 01, \pm 5, \pm 7\} \tag{4.92}$$

Fig. 4.13. (a) gives the comparison of the power components, showing

$$\hbar_{p^*_{ij}} \in \{\pm 2, 02, \pm 4, 04, \pm 6, 06, \pm 8, 08, +10, 010, \pm 12, 012, -14, 014\} \tag{4.93}$$

close to the simulation results. The set of harmonics of the reference capacitor voltages from the proposed calculation is shown as

$$\hbar_{u^*_{Cij}} \in \{\pm 2, 02, \pm 4, 04, \pm 6, 06, \pm 8, 08, \ldots\} \tag{4.94}$$

the accuracy is validated by Table 4.6. Conventionally there exists $\hbar_{u^*_{Cij}} = \hbar_{p^*_{ij}}$.

Fig. 4.13.(b) shows the proposed switching signals

$$\hbar_{s^*_{ij}} \in \{\pm 1, 01, \pm 3, 03, \pm 5, 05, \cdots\} \tag{4.95}$$

which proves its accuracy, while in the conventional method the switching functions satisfy $\hbar_{s^*_{ij}} = \hbar_{u^*_{ij}}$.

It is obvious to see that the waveforms of capacitor voltages and switching functions/modulation signals from the proposed method are closer to the simulated results (Fig. 4.14), indicating its high accuracy, while the waveforms of the conventional calculation show distinct errors.

4.7. Discussion

This chapter proposes the harmonic interaction analysis of the delta-connected CHB as shunt APF. The procedure of the harmonic interaction analysis is explained step by step, resulting in an expression in which the symmetrical components in each order harmonic in the variables can be decoupled while preserving the coupling with variables, which also contain harmonics with sequences. This chapter establishes the power flow harmonics in symmetrical components at the first time, and gives their complete and accurate relation to the capacitor voltages, resulting in the accurate calculation of capacitor voltage harmonics. The ac side dynamic is presented without neglecting the capacitor voltage ripple, leading to the switching signal calculation in a precise way. Compared with the conventional approach which does not consider the harmonic interaction, the proposed calculation of the harmonic capacitor voltages and switching signals is evidently more precise, and shows the capability of solving more existing harmonics in the electrical and non-electrical quantities. The simulation results in MATLAB/Simulink demonstrate the high accuracy of the proposed approach, providing a great reference for the understanding, designing and controlling of the delta-connected CHB-based shunt APF.

With a wide range of power electronics-related applications in power systems, not only harmonic currents are increasing at an alarming rate which has greatly deteriorated the power quality in electrical power networks, but also equipments such as cycloconverter, arc furnaces and fluctuating loads, produce frequencies, which are not an integer of the fundamental frequency. Interharmonics can be generated at and transferred to any voltage level. They can appear as discrete frequencies or a wide band spectrum, severely degrading the power system performance. For instance, it has shown that the general expression for the interharmonic frequencies produced by a cycloconverter with a p_1-pulse converter and a p_2-pulse inverter is as follows:

$$f_{IH} = |(p_1 m \pm 1) f_1 \pm k p_2 f_o| \tag{4.96}$$

where f_{IH} is the interharmonics imposed on the power systems, f_1 and f_o are are the system fundamental and drive frequencies, m is a positive integer including zero (0, 1, 2, $\cdots$), and k is a positive integer (1, 2, 3, $\cdots$). For example, the main interharmonic frequencies of a 6-pulse cycloconverter running at 43 Hz (for the case f_1 =50 Hz) are $|50 \pm 6 \times 43|$ or 208 Hz and 308 Hz.

When there are interharmonics of a wide band spectrum, the analysis technique may not be practical. The application of the proposed harmonic interaction analysis technique can be practical when a single, or a minimal number of, interharmonic frequencies are a concern. The procedure in Sections 4.3-4.4 can be used to analyze frequencies of concern. We can take the interaction of three-phase voltages containing one single interharmonic, the frequency of which is denoted as f_{su}, and three-phase currents containing another single interharmonic, the frequency of which is denoted as f_{si}, as an example.

$$\begin{aligned}\boldsymbol{u}^{abcf_{su}} &= \boldsymbol{u}^{abc+f_{su}} + \boldsymbol{u}^{abc-f_{su}} + \boldsymbol{u}^{abc0f_{su}} \\ &= \boldsymbol{T}^{2r/3s}_{+f_{su}}\boldsymbol{u}^{dq+f_{su}} + \boldsymbol{T}^{2r/3s}_{-f_{su}}\boldsymbol{u}^{dq-f_{su}} + \boldsymbol{T}^{2r/3s}_{0f_{su}}\boldsymbol{u}^{dq0f_{su}}\end{aligned} \tag{4.97}$$

$$\begin{aligned}\boldsymbol{i}^{abcf_{si}} &= \boldsymbol{i}^{abc+f_{si}} + \boldsymbol{i}^{abc-f_{si}} + \boldsymbol{i}^{abc0f_{si}} \\ &= \boldsymbol{T}^{2r/3s}_{+f_{si}}\boldsymbol{i}^{dq+f_{si}} + \boldsymbol{T}^{2r/3s}_{-f_{si}}\boldsymbol{i}^{dq-f_{si}} + \boldsymbol{T}^{2r/3s}_{0f_{si}}\boldsymbol{i}^{dq0f_{si}}\end{aligned} \tag{4.98}$$

Similar with Eqs. (4.27)-(4.28), Eqs. (4.97)-(4.98) indicate that the *abc* frame voltages with interharmonic frequency f_{su} and currents with interharmonic frequency f_{si} can be expressed by the *dq* variables rotating with the corresponding frequency f_{su} and f_{si}. $+, -, 0$ indicate the positive-, negative- and zero-sequence component. $\boldsymbol{T}^{2r/3s}_{\lambda\Xi}$ ($\lambda \in \{+, -, 0\}$, $\Xi \in \{f_{su}, f_{si}\}$) is transformation matrix for transferring two-phase rotating signals to three-phase stationary signals of the frequency Ξ. Define the interaction of the voltage of Eq. (4.97) and the current of Eq. (4.98) as

$$\breve{\boldsymbol{p}}^{abc} = \boldsymbol{u}^{abcf_{su}} \circ \boldsymbol{i}^{abcf_{si}} \tag{4.99}$$

where $\circ$ has the same definition in Section 4.2. It can be expected that the resulted three-phase power flows of Eq. (4.99) include three cases from the analysis procedure introduced in Section 4.3:

- $\breve{\boldsymbol{p}}^{abc}$ contains frequencies $(f_{su} + f_{si})$ and $(f_{su} - f_{si})$ if $f_{su} > f_{si}$. There is $\breve{\boldsymbol{p}}^{abc} = \breve{\boldsymbol{p}}^{abc(f_{su}+f_{si})} + \breve{\boldsymbol{p}}^{abc(f_{su}-f_{si})}$.
- $\breve{\boldsymbol{p}}^{abc}$ contains frequencies $(f_{su} + f_{si})$ and $(f_{si} - f_{su})$ if $f_{su} < f_{si}$. There is $\breve{\boldsymbol{p}}^{abc} = \breve{\boldsymbol{p}}^{abc(f_{su}+f_{si})} + \breve{\boldsymbol{p}}^{abc(f_{si}-f_{su})}$.
- $\breve{\boldsymbol{p}}^{abc}$ contains the frequency $(f_{su} + f_{si})$ and a dc component. There is $\breve{\boldsymbol{p}}^{abc} = \breve{\boldsymbol{p}}^{abc(f_{su}+f_{si})} + \breve{\boldsymbol{p}}^{abc}_{dc}$.

When the three-phase voltages and currents contain a number of interharmonics, the similar analysis procedure introduced in Section 4.4 can be used to derive the *dc* component and the *dq* variables of the *ac* component in the resulted powers.

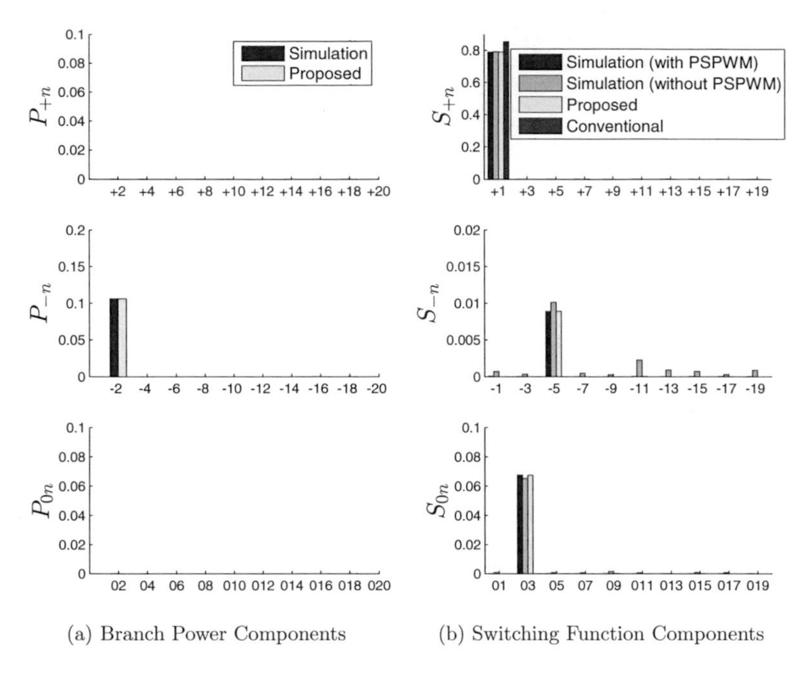

(a) Branch Power Components (b) Switching Function Components

Figure 4.5.: Case A with the PHC strategy (a) harmonic sequence amplitudes of branch powers in pu (base value 25 MVA) and (b) harmonic sequence amplitudes of switching functions in pu.

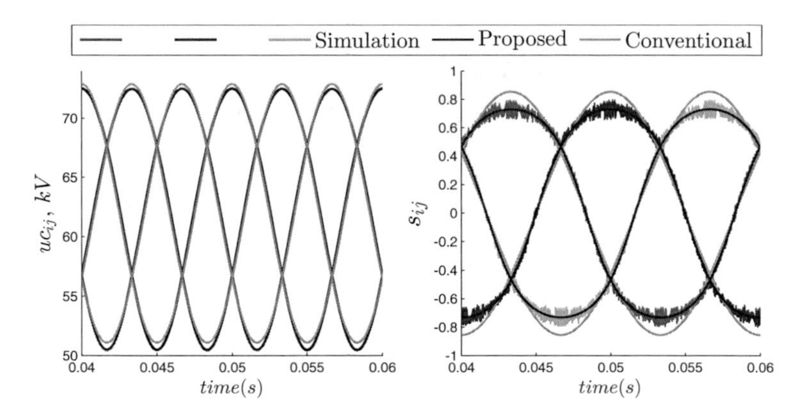

Figure 4.6.: Case A with the PHC strategy: the sum of SM capacitor voltages in each branch (left) and switching functions (right) in time-domain

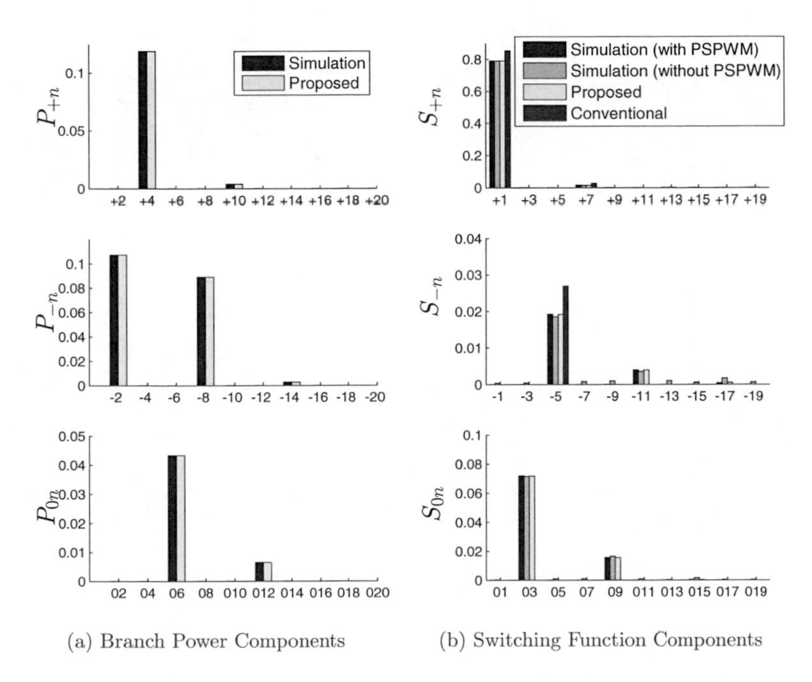

(a) Branch Power Components (b) Switching Function Components

Figure 4.7.: Case B with the PHC strategy (a) harmonic sequence amplitudes of branch powers in pu (base value 25 MVA) and (b) harmonic sequence amplitudes of switching functions in pu.

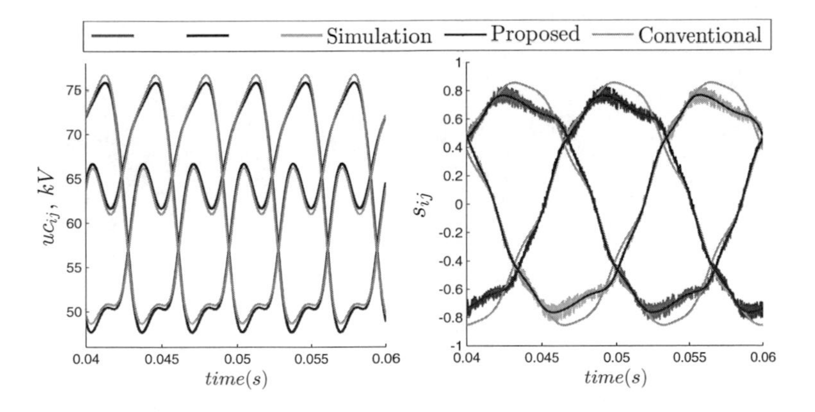

Figure 4.8.: Case B with the PHC strategy: the sum of SM capacitor voltages in each branch (left) and switching functions (right) in time-domain

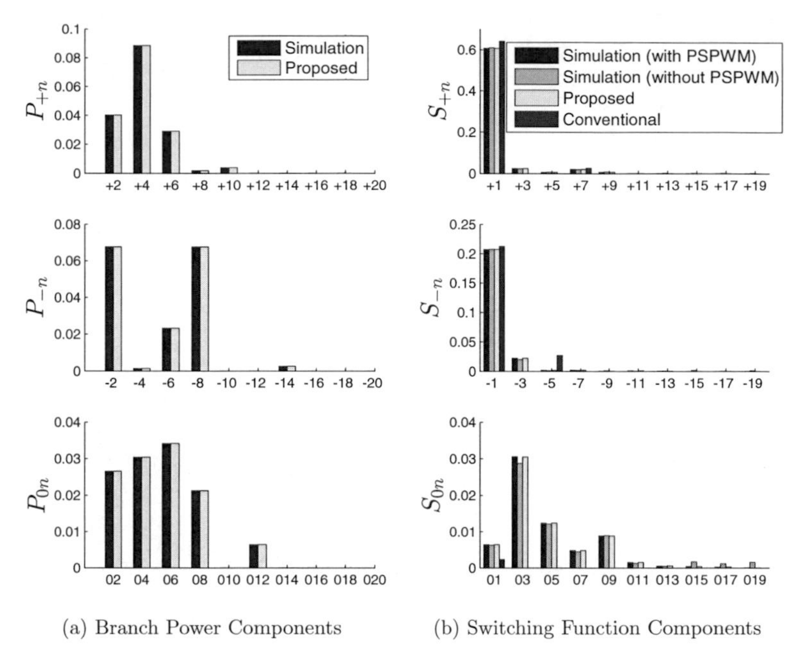

(a) Branch Power Components (b) Switching Function Components

Figure 4.9.: Case C with the PHC strategy (a) harmonic sequence amplitudes of branch powers in pu (base value 25 MVA) and (b) harmonic sequence amplitudes of switching functions in pu.

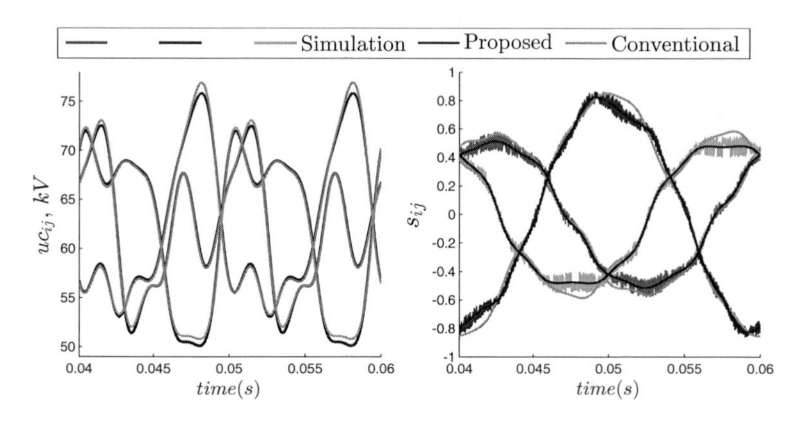

Figure 4.10.: Case C with the PHC strategy: the sum of SM capacitor voltages in each branch (left) and switching functions (right) in time-domain

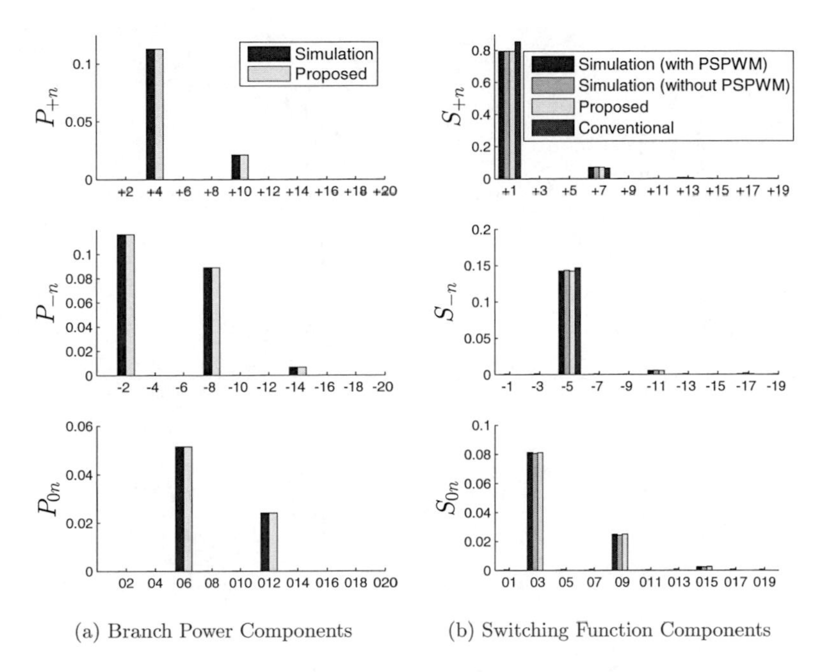

(a) Branch Power Components (b) Switching Function Components

Figure 4.11.: Case D with the PHC strategy (a) harmonic sequence amplitudes of branch powers in pu (base value 25 MVA) and (b) harmonic sequence amplitudes of switching functions in pu.

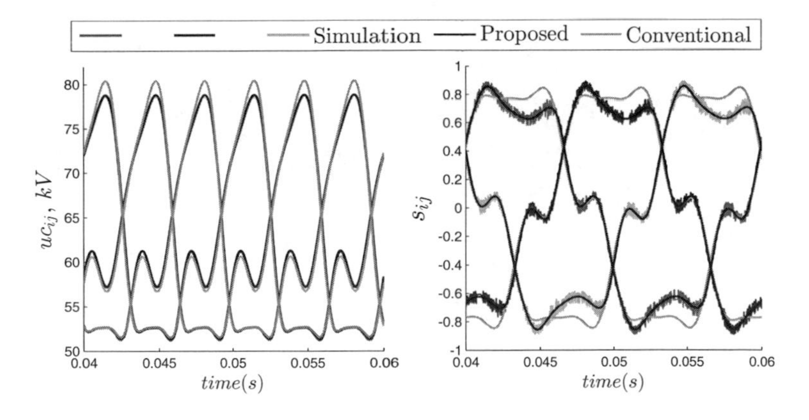

Figure 4.12.: Case D with the PHC strategy: the sum of SM capacitor voltages in each branch (left) and switching functions (right) in time-domain

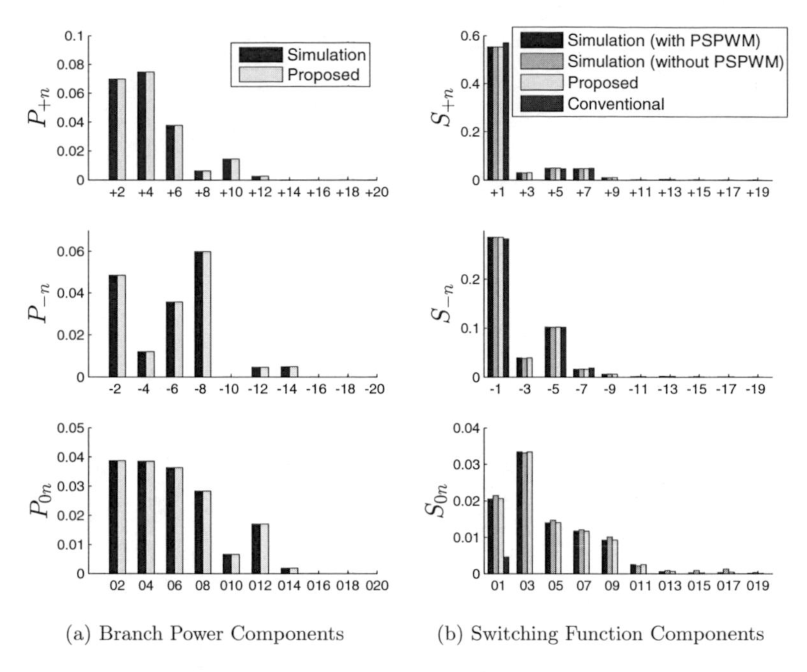

(a) Branch Power Components (b) Switching Function Components

Figure 4.13.: Case E with the PHC strategy (a) harmonic sequence amplitudes of branch powers in pu (base value 25 MVA) and (b) harmonic sequence amplitudes of switching functions in pu.

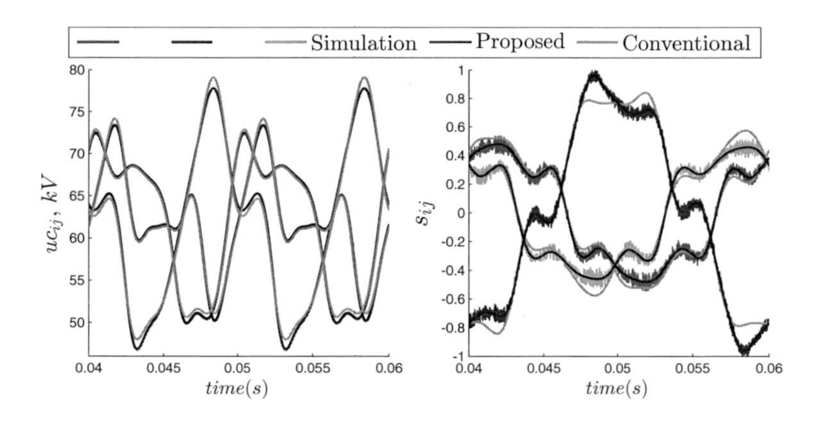

Figure 4.14.: Case E with the PHC strategy: the sum of SM capacitor voltages in each branch (left) and switching function (right) in time-domain

Table 4.2.: Harmonic sequence amplitudes of capacitor sum voltages of Case A

Case	A				
	Simulation (V)	Proposed		Conventional	
		value (V)	error (%)	value (V)	error (%)
U_{C-2}	**10966**	**10944**	0.2	**10900**	0.6
U_{C+4}	**483.2**	**486.7**	0.7	/	/
U_{C06}	**46.3**	**43.4**	6.3	/	/

Table 4.3.: Harmonic sequence amplitudes of capacitor sum voltages of Case B

Case	B				
	Simulation (V)	Proposed		Conventional	
		value (V)	error (%)	value (V)	error (%)
U_{C-2}	**11153**	**11145**	0.1	**11011**	1.3
U_{C+4}	**6127**	**6127**	0	**6106**	0.3
U_{C06}	**1153**	**1145**	0.7	**1482**	28.5
U_{C-8}	**2363**	**2365**	0.1	**2281**	3.5
U_{C+10}	**220**	**215**	2.3	**79**	64.1
U_{C012}	**31**	**32**	3.2	**111**	258.1
U_{C-14}	**51**	**53**	3.9	**39**	23.5
U_{C+16}	**31**	**28**	9.7	/	/

Table 4.4.: Harmonic sequence amplitudes of capacitor sum voltages of Case C

Case	C				
	Simulation (V)	Proposed		Conventional	
		value (V)	error (%)	value (V)	error (%)
U_{C+2}	**4107.9**	**4093.2**	0.36	**4125**	0.41
U_{C-2}	**7009.7**	**6998.9**	0.15	**6946**	0.91
U_{C02}	**2680.7**	**2666.9**	0.51	**2725.1**	1.66
U_{C+4}	**4478.4**	**4472.5**	0.13	**4544.8**	1.48
U_{C-4}	**187.4**	**192.7**	2.83	**68.7**	63.34
U_{C04}	**1591.6**	**1590.5**	0.07	**1560.7**	1.94
U_{C+6}	**922.6**	**920.3**	0.25	**992.4**	7.57
U_{C-6}	**730.0**	**730.4**	0	**797.1**	9.19
U_{C06}	**1004.9**	**1001.9**	0.30	**1170.1**	16.44
U_{C+8}	**124.3**	**124.2**	0	**48.1**	61.30
U_{C-8}	**1735.8**	**1736.1**	0	**1735.0**	0
U_{C08}	**641.2**	**639.6**	0.25	**546.3**	14.80
U_{C+10}	**75.7**	**74.7**	1.32	**79.1**	4.49
U_{C-10}	**86.2**	**87.6**	1.62	/	/
U_{C010}	**53.9**	**53.0**	1.67	/	/
U_{C+12}	**21.2**	**19.8**	6.60	/	/
U_{C-12}	**21.8**	**21.8**	0	/	/
U_{C012}	**46.5**	**45.3**	2.58	**110.8**	138.28
U_{C+14}	**7.4**	**8.8**	18.92	/	/
U_{C-14}	**47.1**	**48.7**	3.40	**38.8**	17.62
U_{C014}	**18.7**	**17.3**	7.49	/	/
U_{C+16}	**14.3**	**13.4**	6.29	/	/
U_{C-16}	**11.9**	**12.5**	5.04	/	/
U_{C016}	**4.4**	**2.1**	52.27	/	/

Table 4.5.: Harmonic sequence amplitudes of capacitor sum voltages of Case D

Case	D				
	Simulation (V)	Proposed		Conventional	
		value (V)	error (%)	value (V)	error (%)
U_{C-2}	**11589**	**11567**	0.2	**11929**	2.9
U_{C+4}	**5316**	**5331**	0.3	**5798**	9.1
U_{C06}	**1442**	**1439**	0.2	**1759**	22.0
U_{C-8}	**2382**	**2382**	0	**2280**	4.3
U_{C+10}	**289**	**288**	0.3	**432**	49.5
U_{C012}	**415**	**416**	0.2	**411**	1.0
U_{C-14}	**156**	**154**	1.3	**97**	37.8
U_{C+16}	**42**	**43**	2.4	/	/

Table 4.6.: Harmonic sequence amplitudes of capacitor sum voltages of Case E

Case	E				
	Simulation (V)	Proposed		Conventional	
		value (V)	error (%)	value (V)	error (%)
U_{C+2}	**7340.6**	**7334.9**	0.07	**7172.1**	2.30
U_{C-2}	**4979.1**	**4960.1**	0.38	**4989.9**	0.22
U_{C02}	**3977.6**	**3959.9**	0.44	**3973.6**	0.10
U_{C+4}	**3685.5**	**3690.5**	0.14	**3839.4**	4.18
U_{C-4}	**917.5**	**917.4**	0	**619.1**	32.52
U_{C04}	**2202.2**	**2208.0**	0.26	**1979.5**	10.11
U_{C+6}	**1298.2**	**1294.0**	0.32	**1293.6**	0.35
U_{C-6}	**1131.5**	**1132.7**	0.11	**1224.9**	8.25
U_{C06}	**1122.2**	**1120.4**	0.16	**1243.4**	10.80
U_{C+8}	**196.3**	**196.1**	0.10	**161.8**	17.58
U_{C-8}	**1460.3**	**1462.3**	0.14	**1537.9**	5.31
U_{C08}	**864.5**	**865.2**	0.08	**727.3**	15.87
U_{C+10}	**318.4**	**317.2**	0.38	**298.2**	6.34
U_{C-10}	**85.1**	**86.2**	1.29	/	/
U_{C010}	**207.9**	**208.0**	0	**136.9**	34.15
U_{C+12}	**31.7**	**31.4**	0.95	**45.6**	43.85
U_{C-12}	**55.3**	**56.1**	1.45	**79.9**	44.48
U_{C012}	**293.2**	**292.9**	0.10	**291.7**	0.51
U_{C+14}	**29.2**	**29.0**	0.68	/	/
U_{C-14}	**82.7**	**84.1**	1.69	**72.5**	12.33
U_{C014}	**47.2**	**46.6**	1.27	**27.4**	41.95
U_{C+16}	**11.2**	**9.2**	17.86	/	/
U_{C-16}	**19.1**	**19.5**	2.09	/	/
U_{C016}	**9.1**	**9.3**	2.20	/	/

5. Summary and Outlook

Nowadays the wide utilisation of power-electronics-based loads cause current harmonic injection and reactive power components at the grids, the electrical power quality has become an important technical issue in medium-voltage transmission and distribution power systems. To address these problems, some grid codes impose requirements to ensure power quality at the grid side, shunt APFs based on the voltage source converter technology have been a proven solution. Delta-connected Cascaded H-bridge (CHB) is one attractive shunt active power filter (APF) configuration since multiple H-bridges cascaded in series enables the converter to be directly connected to medium-voltage grids without the presence of a bulky and lossy line-frequency transformer. In addition, it has many advantages such as modular structure with easy construction and maintenance, less harmonic generation, and low switching losses compared with the tradtional voltage source converters.

In this dissertation the application of the delta-connected CHB multilevel converter as shunt APF under nonideal power supplies is introduced. Two issues have been tackled, one is an optimal operation current strategy to determinine the compensating currents (or called terminal currents) and the circulating current for APF apparent power minimization. Apparent power reduction is an important practical concern since apparent powers are not the only determinant of costs but a very important factor for circuit design. Another issue is harmonic interaction analysis so as to predict how harmonics propagate through the system and quantify the electrical and non-electrical quantities. The dc-link application with film capacitors, which have lower capacitance than electrolytic type capacitors with the same volume, leads to higher amplitude low-frequency ($<2\,\mathrm{kHz}$) capacitor voltage ripple, however, can achieve lower cost and higher reliability. In addition, for the same switch and dc-link capacitor in each SM, the higher number of SMs is accompanied with the higher quality of the output voltage. However, the total number of devices, converter cost, and conduction loss are also higher. Therefore, a minimum number of SMs, and each SM with a highly reliable film capacitor, are desired to reduce the converter volume, cost, losses, and failures in cascaded multilevel converters. Harmonic interaction analysis is very useful at the APF designing and controlling stage. At the APF designing stage such analysis is helpful for circuit parameters design and semiconductor devices selection. At the controlling stage, accurate variable quantification is important for control strategies that are sensitive to model errors.

After a brief introduction of the background, Chapter 1 leads into multilevel converters. The application of various multilevel converter configurations is introduced, including the well-known multilevel converters, for example neutral point clamped converter, flying capacitor converter, and modular multilevel cascade converters. The reasons to select the delta-connected cascaded H-bridge multilevel converter as shunt APF are given. In addition, the mathematical expression of the non-ideal power supplies, which contain harmonics and unbalance is introduced as the foundation of the future chapters.

In Chapter 3 an optimal current operation strategy for a delta-connected CHB-based shunt APF under non-ideal grid conditions is presented that minimizes the APF operational power and satisfies requirements on average power balance, power factor constraint, source current distortion constraint as per IEEE STD-519 and imbalance characteristics as per IEEE STD-1159. This optimal strategy leads to a non-convex Quadratic Constraints Quadratic Programming (QCQP) with a convex objective function and non-convex constraints. Non-convex problems are difficult to solve. All known algorithms to solve them have a complexity that grows exponentially with problem dimensions. As the non-convex QCQP problem is NP-hard, the optimization methods are typically based on convex relaxations of the problem. The well-known relaxation methods, such as SDP and RLT relaxations obtain tighter bounds at the expense of a larger number of variables and constraints, moreover, they needs techniques such as randomization procedure to extract the good feasible solution of the original non-convex QCQP problem from the optimal solution of the relaxed problem. In this dissertation, SCP (sequential convex programming) has been applied to solve the non-convex QCQP problem. By linearising the non-convex constraints at a feasible starting point, the non-convex QCQP problem is transformed to a convex problem which can be solved efficiently, leading to a new feasible point with a lower objective value. If we linearise again the non-convex function around the new feasible point and repeat the procedure, we can obtain a sequence of feasible points with decreasing objective values. The proposed algorithm converges to a local optimal solution mostly within 2 or 3 iterations based on our computer simulations which represent typical scenarios of non-ideal grid voltage and load current conditions.

Chapter 4 proposes harmonic interaction analysis of the delta-connected CHB as shunt APF based on the assumption that the steady-state switching functions/modulation signals are unknown beforehand. A time-invariant *dq* model has been derived from the time-varying model in *abc*-frame, by application of park transformation at symmetrical components of different frequency orders. In the derived time-invariant representation the harmonic sequences in the variables can be decoupled while preserving the coupling with other variables, which also contain harmonics with sequences. This chapter establishes the power flow harmonics in symmetrical components at the first time, and gives their accuare relation to the SM capacitor voltages, resulting in the accurate calcution of capacitor voltage harmonics. The ac side dynamic is presented without neglecting the capacitor voltage ripple, leading to the precise low-frequency switching function calculation which is uniform to any modulation strategy since the switching function harmonics are brought during the interaction with the SM capacitor voltages instead of the modulation stage.

The research proposed in Chapters 3 - 4 can be extended to the following directions:

In Chapter 3 the objective in the optimal current strategy is to minimize the APF apparent power. The proposed strategy can be easily extended for other objectives, for example, branch current selective harmonic minimization/capacitor voltage ripple minimization. It can be expected that such optimization problems are also non-convex QCQP and can be solved by the sequential convex programming.

In Chapter 4 the low-frequency harmonics in switching functions are generated by the interaction with the SM capacitor voltages. To validate the harmonic interaction analysis, the ratio between the carrier frequency and the line-frequency has to be high in order to shift the first sideband to high frequeny (>40th). In this way, the effect of modulation can be ignored. However, when the ratio is low, the low-frequency harmonics in switching functions will be impacted by the modulation. The present research on harmonic interaction analysis with the consideration of modulation impact to switching functions is based on that the steady-state modulation signals contain only the fundamental frequency component, such as in STATCOM applications. However, in active power filter applications, the steady-state modulation signals contain not only the fundamental frequency but also low-frequency harmonics. Therefore, when the ratio between the carrier frequency and the line-frequency is low, the analysis of switching functions in active power filter applications is more complicated than STATCOM applications, and remains an open problem.

6. Zusammenfassung

Die breite Nutzung von leistungselektronikbasierten Lasten verursacht Stromoberschwingungen und Blindleistungen am Netz. Daher ist die Qualität der elektrischen Leistung ist zu einem wichtigen Thema bei Mittelspannungsübertragungs und Verteilungssysteme geworden. Hier ermöglicht die Kompensation von Stromoberschwingungen und Blindleistungen, unter Einsatz von Spannungsquellenwandlerbasierten Aktivleistungsfiltern die Einhaltung von Netzregeln der Übertragungsnetzbetreiber. Der Delta-verbundene Kaskadierende-H-Brücken-Umrichter (CHB) ist eine attraktive Konfiguration dieses Aktivleistungsfilters. Durch die Reihenschaltung von modularen H-Brücken bietet der CHB-Umrichter vorteilhafte Eigenschaften in Bezug auf hohe Spannungsstufen sowie Nennleistungen. Darüber hinaus bietet der Umrichter viele Vorteile, wie z. B. die modulare Struktur mit einfacher Konstruktion und Wartung, veringerten Überschwingungsanteil im Vergleich zu konventionellen Spannungsquellenwandlern.

Die Dissertation widmet sich der Untersuchung von delta-verbundenen CHB für Mittelspannungsanwendungen. Zwei Probleme wurden behandelt: eine optimale Strategie für die Ausgleichsströme und der Kreisstrom zu bestimmen, um die Leistung vom delta-verbundenen CHB-Umrichter zu minimieren. Die Leistung zu reduzieren ist wichtig, da reduzierte Leistungen zu geringeren Kosten führen können.

Das zweite Thema ist die Analyse der harmonischen Interaktion, damit sich die Oberwellen vorhersagen lassen, die sich durch das System ausbreiten, und sich die elektrische und nichtelektrische Variablen quantifizieren. Die dc-Link-Anwendung mit Folienkondensatoren, die eine geringere Kapazität als elektrolytische Kondensatoren bei gleichen Volumen haben, führt zu höherer niederfrequenter (<2 kHz) Kondensatorspannungswelligkeit, jedoch können Folienkondensatoren niedrigere Kosten und höhere Zuverlässigkeit erreichen. Außerdem geht die höhere Anzahl von SMs mit höherer Qualität der Ausgangsspannung für den gleichen Schalter und den Zwischenkreiskondensator in jedem SM einher. Die Gesamtanzahl der Geräte, die Konverterkosten und der Leitungsverlust sind jedoch höher. Daher ist eine minimale Anzahl von SMs und die Verwendung von hochzuverlässigen Filmkondensatoren erwünscht, um das Konvertervolumen, die Kosten, Verluste und Ausfälle in kaskadierten Multilevel-Konvertern zu reduzieren. Die Analyse der harmonischen Interaktion ist in der APF-Entwurfs- und Kontrollphase sehr nützlich. In der APF-Entwurfsphase ist eine solche Analyse für den Entwurf von Schaltungsparametern und die Auswahl von Halbleitervorrichtungen hilfreich. In der Kontrollphase ist eine genaue Quantifizierung von Variablen für Steuerungsstrategien wichtig, die empfindlich gegenüber Modellfehlern sind.

Die Familie der Multilevel-Umrichtern und deren Anwendungen werden in Kapitel 2 vorgestellt. Die Familie umfässt herkömmlichen Multilevel-Topologien, wie Diode-Clamped-Umrichter, Flying-Capacitor-Umrichter und kaskadierenden modularen Multilevel-Umrichter. Außerdem werden die mathematische Beschreibung von nicht-idealen Netzspannungen als die Grundlage für zukünfitige Kapitel eingeführt.

Im Kapitel 3 wird eine optimale Stromstrategie, die die Leistungen von APF minimiert und die Anforderung an Leistungenbalance, Leistungsfaktor, Quellenstromverzerrungsbeschränkung gemäß IEEE Std-519, die Unwucht des Quellstroms gemäß IEEE Std-1159 erfüllt. Die Stromstrategie wird als eine quadratische Optimierung mit einer quadratischen Zielfunktion unter nicht-konvexen quadratischen Nebenbedingungen (QCQP) formuliert. Nicht-konvexe Optimierungen sind schwer zu löschen. Alle bekannten Algorithmen, die QCQP Probleme zu löschen, befinden sich eine Komplexität, die exponentiell wächst mit der Größe der Probleme. Da die QCQP Probleme NP-hart sind, basieren die Optimierungsmethoden typischerweise auf der Relaxierung der Probleme. Die bekannten Relaxierungsmethoden sowie SDP auch RLT erreichen engere Grenzen auf Kosten von einer größeren Anzahl von Variablen und Einschränkungen. Außerdem werden Techniken, z.B., Randomisierungsverfahren gebraucht, um die Lösung des ursprünglichen QCQP Problems von der des relaxierten Problems zu extrahieren. In dieser Dissertation wird die SCP (sequential convex programming) Methode verwendet, um die nicht-konvexe Optimierung zu löschen. Durch Linearisierung der nichtkonvexen Nebenbedingungen an einer zulässigen Startlösung wird die nicht-konvexe QCQP Optimierung in einer konvexen umgewandelt, die effizient gelöst werden kann und zu einer neuen zulässigen Lösung mit einem niedrigeren Zielwert führt. Wenn wir widerum die nichtkonvexe Funktion um den neuen zulässigen Punkt linearisieren und den Verfahren widerholen, können wir eine Abfolge von zulässigen Punkten mit abnehmenden Zielwerten erhalten. Basierend auf dem typischen Szenario nicht idealer Netzspannungs- und Laststrombedingungen konvergiert der präsentierte Algorithmus zu einer lokalen optimalen Lösung meist innerhalb von 2 oder 3 Iterationen.

Im Kapitel 4 wird die Analyse der harmonischen Wechselwirkung vom Delta-verbundenen CHB-basierten Shunt APF mit der Annahme, dass die stationäre Schaltfunktion vorher unbekannt ist, vorgestellt. Ein zeitinvariantes dq-Modell wird aus dem zeitveränderlichen Modell im abc-Rahmen abgleitet, durch Anwendung der Parktransformation der symmetrischen Komponenten unterschiedlicher Frequenzordnungen. In der zeitinvarianten Darstellung können die symmetrischen Komponenten in jeder Frequenzordnung in den Variablen entkoppelt werden, unter Beibehaltung der Kopplung mit anderen Variablen, die Harmonische mit Sequenzen enthalten. In diesem Kapitel werden sowohl die Oberschwingungen des Leistungsflusses in symmetrischen Komponenten als auch das genaue Verhältnis zwischen den Leistungsflüssen und den Kondensatorspannungen festgelegt. Die führen zur genauen Berechnung der Oberwellen der Kondensatorspannung. Die ac-Seite Dynamik wird dargestellt, ohne die Oberwellen der Kondensatorspannung zu vernachlässigen. Dies führt zu einer genauen Berechnung der niederfrequenten Schaltfunktion, was für jede Modulationsstrategie einheitlich ist, da die Schaltfunktionsoberwellen während der Interaktion mit den SM-Kondensatorspannungen anstelle der Modulationsstufe gebracht werden.

Die in Kapiteln 3 - 4 beschreibene Forschung kann auf folgenden Richtungen erweitert werden:

Das Ziel der optimalen Strategie in Kapitel 3 ist die APF-Leistung zu minimieren. Mit Hilfe der Interaktionsanalyse aus Kapitel 4 kann eine optimale Stromstrategie aufgestellt werden, bei der das Ziel die Minimierung von Oberwellen der Kondensatorspannung sein kann, weil die Oberwellen der Kondensatorspannung von der Zweigströme beeinflusst werden. Man kann erwarten, dass das Minimierungsproblem auch ein nicht konvexes QCQP-Problem ist.

Die niederfrequenten Oberwellen in den Schaltfunktionen in Kapitel 4 werden durch die Interaktion mit dem SM-Kondensatorsspannungen erzeugt. Das Verhältnis zwischen der Trägerfrequenz und der Netzfrequenz muss hoch sein, um das erste Seitenband auf eine hohe Frequenz (>40th) zu verschieben. Auf diese Weise kann der Effekt der Modulation ignoriert werden. Wenn das Verhältnis jedoch niedrig ist, werden die niederfrequenten Oberwellen in den Schaltfunktionen durch die Modulation beeinflusst. Bislang basiert die Analyse der Oberschwingungen unter Berücksichtigung der Auswirkung der Modulation auf die Schaltfunktion darauf, dass das stationäre Modulationssignal nur die Grundfrequenz enthält, z.B. in der Anwendung von STATCOM. In Anwendungen von Aktivleistungsfiltern enthält das stationäre Modulationssignal jedoch nicht nur die Grundfrequenz, sondern auch niederfrequente Oberschwingungen. Wenn das Verhältnis zwischen der Trägerfrequenz und der Netzfrequenz niedrig ist, ist die Analyse von Schaltsignalen mit Modulation in der APF-Anwendungen daher komplizierter als der STATCOM-Anwendungen und bleibt ein offenes Problem.

A. Supplementary Materials

A.1. Power Component Calculation (Chapter 3)

$$\boldsymbol{p}^{*abc} = \begin{pmatrix} p_{ab}^* \\ p_{bc}^* \\ p_{ca}^* \end{pmatrix} = \begin{pmatrix} v_{sab} i_{ab}^* \\ v_{sbc} i_{bc}^* \\ v_{sca} i_{ca}^* \end{pmatrix} \tag{A.1.1}$$

$$\boldsymbol{q}^{*abc} = \begin{pmatrix} q_{ab}^* \\ q_{bc}^* \\ q_{ca}^* \end{pmatrix} = \frac{1}{\sqrt{3}} \begin{pmatrix} (v_{sbc} - v_{sca}) i_{ab}^* \\ (v_{sca} - v_{sab}) i_{bc}^* \\ (v_{sab} - v_{sbc}) i_{ca}^* \end{pmatrix} \tag{A.1.2}$$

Define

$$\boldsymbol{i}_{ll}^{*abc} = \begin{pmatrix} i_{ab}^* \\ i_{bc}^* \\ i_{ca}^* \end{pmatrix}, \quad \boldsymbol{v}_{sll}^{abc} = \begin{pmatrix} v_{sab} \\ v_{sbc} \\ v_{sca} \end{pmatrix} \tag{A.1.3}$$

$$\boldsymbol{v}_{sllll}^{abc} = \begin{pmatrix} v_{sbc} - v_{sca} \\ v_{sca} - v_{sab} \\ v_{sab} - v_{sbc} \end{pmatrix} = \begin{pmatrix} 0 & 1 & -1 \\ -1 & 0 & 1 \\ 1 & -1 & 0 \end{pmatrix} \boldsymbol{v}_{sll}^{abc} \tag{A.1.4}$$

The dq transformation leads to

$$\boldsymbol{v}_{sllll}^{dq+n} = \begin{pmatrix} v_{sllll,d+n} \\ v_{sllll,q+n} \end{pmatrix} = \begin{pmatrix} \frac{3}{2} & -\frac{\sqrt{3}}{2} \\ \frac{\sqrt{3}}{2} & \frac{3}{2} \end{pmatrix} \boldsymbol{v}_{sll}^{dq+n} \tag{A.1.5}$$

$$\boldsymbol{v}_{sllll}^{dq-n} = \begin{pmatrix} v_{sllll,d-n} \\ v_{sllll,q-n} \end{pmatrix} = \begin{pmatrix} \frac{3}{2} & \frac{\sqrt{3}}{2} \\ -\frac{\sqrt{3}}{2} & \frac{3}{2} \end{pmatrix} \boldsymbol{v}_{sll}^{dq-n} \tag{A.1.6}$$

$$\boldsymbol{v}_{sllll}^{dq0n} = \begin{pmatrix} v_{sllll,d0n} \\ v_{sllll,q0n} \end{pmatrix} = \begin{pmatrix} 1 & 0 \\ 0 & 1 \end{pmatrix} \boldsymbol{v}_{sll}^{dq0n} = \begin{pmatrix} 0 \\ 0 \end{pmatrix} \tag{A.1.7}$$

It can be easily obtained that the sequences in $\boldsymbol{v}_{slll}^{abcn}$ are the same in $\boldsymbol{v}_{sll}^{abcn}$, like for instance, if $\boldsymbol{v}_{sll}^{abcn}$ containes only the positive-sequence component, then $\boldsymbol{v}_{slll}^{abcn}$ will also include the positive-sequence component. Define

$$\boldsymbol{v}_{slll}^{dqn} = \begin{pmatrix} \boldsymbol{v}_{slll}^{dq+n\top} & \boldsymbol{v}_{slll}^{dq-n\top} & \boldsymbol{v}_{slll}^{dq0n\top} \end{pmatrix}^\top \tag{A.1.8}$$

$$\boldsymbol{i}_{ll}^{*dqn} = \begin{pmatrix} \boldsymbol{i}_{ll}^{*dq+n\top} & \boldsymbol{i}_{ll}^{*dq-n\top} & \boldsymbol{i}_{ll}^{*dq0n\top} \end{pmatrix}^\top \tag{A.1.9}$$

$$\boldsymbol{v}_{slll}^{dq} = \begin{pmatrix} \boldsymbol{v}_{slll}^{dq1\top} & \boldsymbol{v}_{slll}^{dq2\top} & \cdots & \boldsymbol{v}_{slll}^{dqM\top} \end{pmatrix}^\top \tag{A.1.10}$$

$$\boldsymbol{i}_{ll}^{*dq} = \begin{pmatrix} \boldsymbol{i}_{ll}^{*dq1\top} & \boldsymbol{i}_{ll}^{*dq2\top} & \cdots & \boldsymbol{i}_{ll}^{*dqM\top} \end{pmatrix}^\top \tag{A.1.11}$$

Average active and reactive powers The APF average active and reactive powers can be calculated as

$$p_{APF,dc}^{*} = \sum_{n=1}^{M} \left(\boldsymbol{v}_{sll}^{dqn\top} \boldsymbol{i}_{ll}^{*dqn} \right) = \boldsymbol{v}_{sll}^{dq\top} \boldsymbol{i}_{ll}^{*dq} \tag{A.1.12}$$

$$q_{APF,dc}^{*} = \frac{1}{\sqrt{3}} \sum_{n=1}^{M} \left(\boldsymbol{v}_{slll}^{dqn\top} \boldsymbol{i}_{ll}^{*dqn} \right) = \frac{1}{\sqrt{3}} \boldsymbol{v}_{slll}^{dq\top} \boldsymbol{i}_{ll}^{*dq} \tag{A.1.13}$$

The calculation results of the PHC, UPF, CST1 and OPT strategies can be seen in Tables 3.7-3.8.

Oscillating active power components Define

$$\boldsymbol{p}^{*dqn} = \begin{pmatrix} \boldsymbol{p}^{*dq+n\top} & \boldsymbol{p}^{*dq-n\top} & \boldsymbol{p}^{*dq0n\top} \end{pmatrix}^\top \tag{A.1.14}$$

$$\boldsymbol{p}^{*dq} = \begin{pmatrix} \boldsymbol{p}^{*dq1\top} & \boldsymbol{p}^{*dq2\top} & \cdots & \boldsymbol{p}^{*dqM\top} \end{pmatrix}^\top \tag{A.1.15}$$

where $\boldsymbol{p}^{*dq}$ collects the dq values of each frequency in $\boldsymbol{p}^{*abc}$ of Eq. (A.1.1). As the derivation process in Chapter 4, $\boldsymbol{p}^{*dq}$ can be calculated as

$$\boldsymbol{p}^{*dq} = \boldsymbol{\mathcal{F}}(\boldsymbol{v}_{sll}^{dq}) \boldsymbol{i}_{ll}^{*dq} \tag{A.1.16}$$

where $\boldsymbol{\mathcal{F}}(\boldsymbol{v}_{sll}^{dq})$ can be otained by replacing $\boldsymbol{u}^{dq}$ in $\boldsymbol{\mathcal{F}}(\boldsymbol{u}^{dq})$ (seen in Eq. (4.48)) by $\boldsymbol{v}_{sll}^{dq}$. It should be clear that the zero-sequence power in this dissertation is not the same meaning of zero-sequence power defined in the instantaneous power theory. In the instantaneous power theory, the zero-sequence power is the product of zero-sequence voltage and zero-sequence current. In this dissertation, the zero-sequence power is, from the point of view of the symmetrical component method, the power flow component which is equal in magnitude and phase in the three phases.

The oscillating power components in Table 3.7- 3.8 are calculated as following,

$$P_{en} = \sqrt{\boldsymbol{p}^{*dqn\top}\text{blkdiag}\left(\frac{1}{3},\frac{1}{3},\frac{1}{3},\frac{1}{3},\frac{1}{2},\frac{1}{2}\right)\boldsymbol{p}^{*dqn}} \tag{A.1.17}$$

Oscillating reactive power components Define

$$\boldsymbol{q}^{*dqn} = \begin{pmatrix}\boldsymbol{q}^{*dq+n\top} & \boldsymbol{q}^{*dq-n\top} & \boldsymbol{q}^{*dq0n\top}\end{pmatrix}^\top \tag{A.1.18}$$

$$\boldsymbol{q}^{*dq} = \begin{pmatrix}\boldsymbol{q}^{*dq1\top} & \boldsymbol{q}^{*dq2\top} & \cdots & \boldsymbol{q}^{*dqM\top}\end{pmatrix}^\top \tag{A.1.19}$$

where $\boldsymbol{q}^{*dq}$ collects the dq values of each frequency in $\boldsymbol{q}^{*abc}$ of Eq. (A.1.2). $\boldsymbol{q}^{*dq}$ can be calculated as

$$\boldsymbol{q}^{*dq} = \frac{1}{\sqrt{3}}\boldsymbol{\mathcal{F}}(\boldsymbol{v}^{dq}_{sllll})\boldsymbol{i}^{*dq}_{ll} \tag{A.1.20}$$

where $\boldsymbol{\mathcal{F}}(\boldsymbol{v}^{dq}_{sllll})$ can be otained by replacing $\boldsymbol{u}^{dq}$ in $\boldsymbol{\mathcal{F}}(\boldsymbol{u}^{dq})$ (seen in Eq. (4.48)) by $\boldsymbol{v}^{dq}_{sllll}$. The oscillating power components in Table 3.7- 3.8 are calculated as following,

$$Q_{en} = \sqrt{\boldsymbol{q}^{*dqn\top}\text{blkdiag}\left(\frac{1}{3},\frac{1}{3},\frac{1}{3},\frac{1}{3},\frac{1}{2},\frac{1}{2}\right)\boldsymbol{q}^{*dqn}} \tag{A.1.21}$$

A discussion about the calculation in Tables 3.7 - 3.8 As Eq. (3.88) and Eq. (3.89) shown,

$$\widetilde{p}^*_{APF} = (p^*_{ab} + p^*_{bc} + p^*_{ca})_{ac} \tag{A.1.22}$$

$$\widetilde{q}^*_{APF} = (q^*_{ab} + q^*_{bc} + q^*_{ca})_{ac} \tag{A.1.23}$$

Therefore $\widetilde{p}^*_{APF}$ and $\widetilde{q}^*_{APF}$, from the aspect of symmetrical component method, are zero-sequence powers, since the summation of the positive sequence of three-phase system is 0, so as the summation of negative sequence. For example, Case A PHC, after calculation, there is $\{-2,+4,06,-8\}$ frequency order components (proved in Chapter 4) in p^*_{ij} and q^*_{ij} , the effective value denoted by respectively P_{en} and Q_{en} $(n = 2,4,6,8)$ shown in Table 3.7. Thus $\widetilde{p}^*_{APF}$ and $\widetilde{q}^*_{APF}$ contains only 6th order component in this case according to the instantaneous power theory, the effective value of $\widetilde{p}^{*}_{APF}$ and $\widetilde{q}^{*}_{APF}$, denoted by $\widetilde{p}^{*}_{APF,e}$ and $\widetilde{q}^{*}_{APF,e}$ (seen in Table 3.3), should be $3P_{e6}$ and $3Q_{e6}$ respectively for Case A PHC.

A.2. Some Details of Eq. (4.26)

Define the nonitalic bold letters as complex space vectors $\mathbf{u}^{dq+m} = u_{d+m} + j \cdot u_{q+m}$, $\mathbf{u}^{dq-m} = u_{d-m} + j \cdot u_{q-m}$, and $\mathbf{u}^{dq0m} = u_{d0m} + j \cdot u_{q0m}$; $\mathbf{i}^{dq+k} = i_{d+k} + j \cdot i_{q+k}$, $\mathbf{i}^{dq-k} = i_{d-k} + j \cdot i_{q-k}$, and $\mathbf{i}^{dq0k} = i_{d0k} + j \cdot i_{q0k}$. A bar above the complex space vectors represents conjugation.

The symmetrical components in Eqs. (4.27)can also be written as the sum of complex conjugated terms,

$$\boldsymbol{u}^{abc+m} = \sqrt{\frac{2}{3}} \cdot \frac{1}{2} \begin{bmatrix} \mathbf{u}^{dq+m} e^{jm\omega t} + \bar{\mathbf{u}}^{dq+m} e^{-jm\omega t} \\ \mathbf{u}^{dq+m} e^{j\left(m\omega t - \frac{2\pi}{3}\right)} + \bar{\mathbf{u}}^{dq+m} e^{-j\left(m\omega t - \frac{2\pi}{3}\right)} \\ \mathbf{u}^{dq+m} e^{j\left(m\omega t + \frac{2\pi}{3}\right)} + \bar{\mathbf{u}}^{dq+m} e^{-j\left(m\omega t + \frac{2\pi}{3}\right)} \end{bmatrix} \tag{A.2.1}$$

$$\boldsymbol{u}^{abc-m} = \sqrt{\frac{2}{3}} \cdot \frac{1}{2} \begin{bmatrix} \mathbf{u}^{dq-m} e^{jm\omega t} + \bar{\mathbf{u}}^{dq-m} e^{-jm\omega t} \\ \mathbf{u}^{dq-m} e^{j\left(m\omega t + \frac{2\pi}{3}\right)} + \bar{\mathbf{u}}^{dq-m} e^{-j\left(m\omega t + \frac{2\pi}{3}\right)} \\ \mathbf{u}^{dq-m} e^{j\left(m\omega t - \frac{2\pi}{3}\right)} + \bar{\mathbf{u}}^{dq-m} e^{-j\left(m\omega t - \frac{2\pi}{3}\right)} \end{bmatrix} \tag{A.2.2}$$

$$\boldsymbol{u}^{abc0m} = \frac{1}{2} \begin{bmatrix} \mathbf{u}^{dq0m} e^{jm\omega t} + \bar{\mathbf{u}}^{dq0m} e^{-jm\omega t} \\ \mathbf{u}^{dq0m} e^{jm\omega t} + \bar{\mathbf{u}}^{dq0m} e^{-jm\omega t} \\ \mathbf{u}^{dq0m} e^{jm\omega t} + \bar{\mathbf{u}}^{dq0m} e^{-jm\omega t} \end{bmatrix} \tag{A.2.3}$$

Similarly Eq. (4.28) can be also rewritten as the sum of complex conjugated terms. Simple multiplication leads in

$$\begin{aligned} \breve{p}_{ab} &= (u_{ab+m} + u_{ab-m} + u_{ab0m})(i_{ab+k} + i_{ab-k} + i_{ab0k}) \\ &= \sqrt{\frac{2}{3}} \cdot \frac{1}{2} \left\{ \left(\frac{1}{2} \mathbf{u}^{dq0m} \mathbf{i}^{dq+k} + \frac{1}{\sqrt{6}} \mathbf{u}^{dq-m} \mathbf{i}^{dq-k} + \frac{1}{2} \mathbf{u}^{dq+m} \mathbf{i}^{dq0k} \right) e^{j((m+k)\omega t)} \right. && ① \\ &\qquad \left. + \left(\frac{1}{2} \bar{\mathbf{u}}^{dq0m} \bar{\mathbf{i}}^{dq+k} + \frac{1}{\sqrt{6}} \bar{\mathbf{u}}^{dq-m} \bar{\mathbf{i}}^{dq-k} + \frac{1}{2} \bar{\mathbf{u}}^{dq+m} \bar{\mathbf{i}}^{dq0k} \right) e^{-j((m+k)\omega t)} \right\} && ② \\ &+ \sqrt{\frac{2}{3}} \cdot \frac{1}{2} \left\{ \left(\frac{1}{\sqrt{6}} \mathbf{u}^{dq+m} \mathbf{i}^{dq+k} + \frac{1}{2} \mathbf{u}^{dq0m} \mathbf{i}^{dq-k} + \frac{1}{2} \mathbf{u}^{dq-m} \mathbf{i}^{dq0k} \right) e^{j((m+k)\omega t)} \right. && ③ \\ &\qquad \left. + \left(\frac{1}{\sqrt{6}} \bar{\mathbf{u}}^{dq+m} \bar{\mathbf{i}}^{dq+k} + \frac{1}{2} \bar{\mathbf{u}}^{dq0m} \bar{\mathbf{i}}^{dq-k} + \frac{1}{2} \bar{\mathbf{u}}^{dq-m} \bar{\mathbf{i}}^{dq0k} \right) e^{-j((m+k)\omega t)} \right\} && ④ \end{aligned}$$

$$
\begin{aligned}
&+\frac{1}{2}\left\{\left(\frac{1}{3}\mathbf{u}^{dq-m}\mathbf{i}^{dq+k}+\frac{1}{3}\mathbf{u}^{dq+m}\mathbf{i}^{dq-k}+\frac{1}{2}\mathbf{u}^{dq0m}\mathbf{i}^{dq0k}\right)e^{j((m+k)\omega t)}\right. && ⑤\\
&\qquad\left.+\left(\frac{1}{3}\bar{\mathbf{u}}^{dq-m}\bar{\mathbf{i}}^{dq+k}+\frac{1}{3}\bar{\mathbf{u}}^{dq+m}\bar{\mathbf{i}}^{dq-k}+\frac{1}{2}\bar{\mathbf{u}}^{dq0m}\bar{\mathbf{i}}^{dq0k}\right)e^{-j((m+k)\omega t)}\right\} && ⑥\\
&+\sqrt{\frac{2}{3}}\cdot\frac{1}{2}\left\{\left(\frac{1}{\sqrt{6}}\mathbf{u}^{dq-m}\bar{\mathbf{i}}^{dq+k}+\frac{1}{2}\mathbf{u}^{dq0m}\bar{\mathbf{i}}^{dq-k}+\frac{1}{2}\mathbf{u}^{dq+m}\bar{\mathbf{i}}^{dq0k}\right)e^{j((m-k)\omega t)}\right. && ⑦\\
&\qquad\left.+\left(\frac{1}{\sqrt{6}}\bar{\mathbf{u}}^{dq-m}\mathbf{i}^{dq+k}+\frac{1}{2}\bar{\mathbf{u}}^{dq0m}\mathbf{i}^{dq-k}+\frac{1}{2}\bar{\mathbf{u}}^{dq+m}\mathbf{i}^{dq0k}\right)e^{-j((m-k)\omega t)}\right\} && ⑧\\
&+\sqrt{\frac{2}{3}}\cdot\frac{1}{2}\left\{\left(\frac{1}{2}\mathbf{u}^{dq0m}\bar{\mathbf{i}}^{dq+k}+\frac{1}{\sqrt{6}}\mathbf{u}^{dq+m}\bar{\mathbf{i}}^{dq-k}+\frac{1}{2}\mathbf{u}^{dq-m}\bar{\mathbf{i}}^{dq0k}\right)e^{j((m-k)\omega t)}\right. && ⑨\\
&\qquad\left.+\left(\frac{1}{2}\bar{\mathbf{u}}^{dq0m}\mathbf{i}^{dq+k}+\frac{1}{\sqrt{6}}\bar{\mathbf{u}}^{dq+m}\mathbf{i}^{dq-k}+\frac{1}{2}\bar{\mathbf{u}}^{dq-m}\mathbf{i}^{dq0k}\right)e^{-j((m-k)\omega t)}\right\} && ⑩\\
&+\frac{1}{2}\left\{\left(\frac{1}{3}\mathbf{u}^{dq+m}\bar{\mathbf{i}}^{dq+k}+\frac{1}{3}\mathbf{u}^{dq-m}\bar{\mathbf{i}}^{dq-k}+\frac{1}{2}\mathbf{u}^{dq0m}\bar{\mathbf{i}}^{dq0k}\right)e^{j((m-k)\omega t)}\right. && ⑪\\
&\qquad\left.+\left(\frac{1}{3}\bar{\mathbf{u}}^{dq+m}\mathbf{i}^{dq+k}+\frac{1}{3}\bar{\mathbf{u}}^{dq-m}\mathbf{i}^{dq-k}+\frac{1}{2}\bar{\mathbf{u}}^{dq0m}\mathbf{i}^{dq0k}\right)e^{-j((m-k)\omega t)}\right\} && ⑫
\end{aligned}
$$

$$
\begin{aligned}
\breve{p}_{bc} &= (u_{bc+m}+u_{bc-m}+u_{bc0m})(i_{bc+k}+i_{bc-k}+i_{bc0k})\\
&=\sqrt{\frac{2}{3}}\cdot\frac{1}{2}\left\{\left(\frac{1}{2}\mathbf{u}^{dq0m}\mathbf{i}^{dq+k}+\frac{1}{\sqrt{6}}\mathbf{u}^{dq-m}\mathbf{i}^{dq-k}+\frac{1}{2}\mathbf{u}^{dq+m}\mathbf{i}^{dq0k}\right)e^{j\left((m+k)\omega t-\frac{2\pi}{3}\right)}\right. && ①\\
&\qquad\left.+\left(\frac{1}{2}\bar{\mathbf{u}}^{dq0m}\bar{\mathbf{i}}^{dq+k}+\frac{1}{\sqrt{6}}\bar{\mathbf{u}}^{dq-m}\bar{\mathbf{i}}^{dq-k}+\frac{1}{2}\bar{\mathbf{u}}^{dq+m}\bar{\mathbf{i}}^{dq0k}\right)e^{-j\left((m+k)\omega t-\frac{2\pi}{3}\right)}\right\} && ②\\
&+\sqrt{\frac{2}{3}}\cdot\frac{1}{2}\left\{\left(\frac{1}{\sqrt{6}}\mathbf{u}^{dq+m}\mathbf{i}^{dq+k}+\frac{1}{2}\mathbf{u}^{dq0m}\mathbf{i}^{dq-k}+\frac{1}{2}\mathbf{u}^{dq-m}\mathbf{i}^{dq0k}\right)e^{j\left((m+k)\omega t+\frac{2\pi}{3}\right)}\right. && ③\\
&\qquad\left.+\left(\frac{1}{\sqrt{6}}\bar{\mathbf{u}}^{dq+m}\bar{\mathbf{i}}^{dq+k}+\frac{1}{2}\bar{\mathbf{u}}^{dq0m}\bar{\mathbf{i}}^{dq-k}+\frac{1}{2}\bar{\mathbf{u}}^{dq-m}\bar{\mathbf{i}}^{dq0k}\right)e^{-j\left((m+k)\omega t+\frac{2\pi}{3}\right)}\right\} && ④\\
&+\frac{1}{2}\left\{\left(\frac{1}{3}\mathbf{u}^{dq-m}\mathbf{i}^{dq+k}+\frac{1}{3}\mathbf{u}^{dq+m}\mathbf{i}^{dq-k}+\frac{1}{2}\mathbf{u}^{dq0m}\mathbf{i}^{dq0k}\right)e^{j((m+k)\omega t)}\right. && ⑤\\
&\qquad\left.+\left(\frac{1}{3}\bar{\mathbf{u}}^{dq-m}\bar{\mathbf{i}}^{dq+k}+\frac{1}{3}\bar{\mathbf{u}}^{dq+m}\bar{\mathbf{i}}^{dq-k}+\frac{1}{2}\bar{\mathbf{u}}^{dq0m}\bar{\mathbf{i}}^{dq0k}\right)e^{-j((m+k)\omega t)}\right\} && ⑥\\
&+\sqrt{\frac{2}{3}}\cdot\frac{1}{2}\left\{\left(\frac{1}{\sqrt{6}}\mathbf{u}^{dq-m}\bar{\mathbf{i}}^{dq+k}+\frac{1}{2}\mathbf{u}^{dq0m}\bar{\mathbf{i}}^{dq-k}+\frac{1}{2}\mathbf{u}^{dq+m}\bar{\mathbf{i}}^{dq0k}\right)e^{j\left((m-k)\omega t-\frac{2\pi}{3}\right)}\right. && ⑦\\
&\qquad\left.+\left(\frac{1}{\sqrt{6}}\bar{\mathbf{u}}^{dq-m}\mathbf{i}^{dq+k}+\frac{1}{2}\bar{\mathbf{u}}^{dq0m}\mathbf{i}^{dq-k}+\frac{1}{2}\bar{\mathbf{u}}^{dq+m}\mathbf{i}^{dq0k}\right)e^{-j\left((m-k)\omega t-\frac{2\pi}{3}\right)}\right\} && ⑧\\
&+\sqrt{\frac{2}{3}}\cdot\frac{1}{2}\left\{\left(\frac{1}{2}\mathbf{u}^{dq0m}\bar{\mathbf{i}}^{dq+k}+\frac{1}{\sqrt{6}}\mathbf{u}^{dq+m}\bar{\mathbf{i}}^{dq-k}+\frac{1}{2}\mathbf{u}^{dq-m}\bar{\mathbf{i}}^{dq0k}\right)e^{j\left((m-k)\omega t+\frac{2\pi}{3}\right)}\right. && ⑨\\
&\qquad\left.+\left(\frac{1}{2}\bar{\mathbf{u}}^{dq0m}\mathbf{i}^{dq+k}+\frac{1}{\sqrt{6}}\bar{\mathbf{u}}^{dq+m}\mathbf{i}^{dq-k}+\frac{1}{2}\bar{\mathbf{u}}^{dq-m}\mathbf{i}^{dq0k}\right)e^{-j\left((m-k)\omega t+\frac{2\pi}{3}\right)}\right\} && ⑩
\end{aligned}
$$

$$
\begin{aligned}
&+\frac{1}{2}\left\{\left(\frac{1}{3}\mathbf{u}^{dq+m}\bar{\mathbf{i}}^{dq+k}+\frac{1}{3}\mathbf{u}^{dq-m}\bar{\mathbf{i}}^{dq-k}+\frac{1}{2}\mathbf{u}^{dq0m}\bar{\mathbf{i}}^{dq0k}\right)e^{j((m-k)\omega t)}\right. && ⑪\\
&\qquad\left.+\left(\frac{1}{3}\bar{\mathbf{u}}^{dq+m}\mathbf{i}^{dq+k}+\frac{1}{3}\bar{\mathbf{u}}^{dq-m}\mathbf{i}^{dq-k}+\frac{1}{2}\bar{\mathbf{u}}^{dq0m}\mathbf{i}^{dq0k}\right)e^{-j((m-k)\omega t)}\right\} && ⑫
\end{aligned}
$$

$$
\begin{aligned}
\breve{p}_{ca} &= (u_{ca+m}+u_{ca-m}+u_{ca0m})(i_{ca+k}+i_{ca-k}+i_{ca0k})\\
&=\sqrt{\frac{2}{3}}\cdot\frac{1}{2}\left\{\left(\frac{1}{2}\mathbf{u}^{dq0m}\mathbf{i}^{dq+k}+\frac{1}{\sqrt{6}}\mathbf{u}^{dq-m}\mathbf{i}^{dq-k}+\frac{1}{2}\mathbf{u}^{dq+m}\mathbf{i}^{dq0k}\right)e^{j\left((m+k)\omega t+\frac{2\pi}{3}\right)}\right. && ①\\
&\qquad\left.+\left(\frac{1}{2}\bar{\mathbf{u}}^{dq0m}\bar{\mathbf{i}}^{dq+k}+\frac{1}{\sqrt{6}}\bar{\mathbf{u}}^{dq-m}\bar{\mathbf{i}}^{dq-k}+\frac{1}{2}\bar{\mathbf{u}}^{dq+m}\bar{\mathbf{i}}^{dq0k}\right)e^{-j\left((m+k)\omega t+\frac{2\pi}{3}\right)}\right\} && ②\\
&+\sqrt{\frac{2}{3}}\cdot\frac{1}{2}\left\{\left(\frac{1}{\sqrt{6}}\mathbf{u}^{dq+m}\mathbf{i}^{dq+k}+\frac{1}{2}\mathbf{u}^{dq0m}\mathbf{i}^{dq-k}+\frac{1}{2}\mathbf{u}^{dq-m}\mathbf{i}^{dq0k}\right)e^{j\left((m+k)\omega t-\frac{2\pi}{3}\right)}\right. && ③\\
&\qquad\left.+\left(\frac{1}{\sqrt{6}}\bar{\mathbf{u}}^{dq+m}\bar{\mathbf{i}}^{dq+k}+\frac{1}{2}\bar{\mathbf{u}}^{dq0m}\bar{\mathbf{i}}^{dq-k}+\frac{1}{2}\bar{\mathbf{u}}^{dq-m}\bar{\mathbf{i}}^{dq0k}\right)e^{-j\left((m+k)\omega t-\frac{2\pi}{3}\right)}\right\} && ④\\
&+\frac{1}{2}\left\{\left(\frac{1}{3}\mathbf{u}^{dq-m}\mathbf{i}^{dq+k}+\frac{1}{3}\mathbf{u}^{dq+m}\mathbf{i}^{dq-k}+\frac{1}{2}\mathbf{u}^{dq0m}\mathbf{i}^{dq0k}\right)e^{j((m+k)\omega t)}\right. && ⑤\\
&\qquad\left.+\left(\frac{1}{3}\bar{\mathbf{u}}^{dq-m}\bar{\mathbf{i}}^{dq+k}+\frac{1}{3}\bar{\mathbf{u}}^{dq+m}\bar{\mathbf{i}}^{dq-k}+\frac{1}{2}\bar{\mathbf{u}}^{dq0m}\bar{\mathbf{i}}^{dq0k}\right)e^{-j((m+k)\omega t)}\right\} && ⑥\\
&+\sqrt{\frac{2}{3}}\cdot\frac{1}{2}\left\{\left(\frac{1}{\sqrt{6}}\mathbf{u}^{dq-m}\bar{\mathbf{i}}^{dq+k}+\frac{1}{2}\mathbf{u}^{dq0m}\bar{\mathbf{i}}^{dq-k}+\frac{1}{2}\mathbf{u}^{dq+m}\bar{\mathbf{i}}^{dq0k}\right)e^{j\left((m-k)\omega t+\frac{2\pi}{3}\right)}\right. && ⑦\\
&\qquad\left.+\left(\frac{1}{\sqrt{6}}\bar{\mathbf{u}}^{dq-m}\mathbf{i}^{dq+k}+\frac{1}{2}\bar{\mathbf{u}}^{dq0m}\mathbf{i}^{dq-k}+\frac{1}{2}\bar{\mathbf{u}}^{dq+m}\mathbf{i}^{dq0k}\right)e^{-j\left((m-k)\omega t+\frac{2\pi}{3}\right)}\right\} && ⑧\\
&+\sqrt{\frac{2}{3}}\cdot\frac{1}{2}\left\{\left(\frac{1}{2}\mathbf{u}^{dq0m}\bar{\mathbf{i}}^{dq+k}+\frac{1}{\sqrt{6}}\mathbf{u}^{dq+m}\bar{\mathbf{i}}^{dq-k}+\frac{1}{2}\mathbf{u}^{dq-m}\bar{\mathbf{i}}^{dq0k}\right)e^{j\left((m-k)\omega t-\frac{2\pi}{3}\right)}\right. && ⑨\\
&\qquad\left.+\left(\frac{1}{2}\bar{\mathbf{u}}^{dq0m}\mathbf{i}^{dq+k}+\frac{1}{\sqrt{6}}\bar{\mathbf{u}}^{dq+m}\mathbf{i}^{dq-k}+\frac{1}{2}\bar{\mathbf{u}}^{dq-m}\mathbf{i}^{dq0k}\right)e^{-j\left((m-k)\omega t-\frac{2\pi}{3}\right)}\right\} && ⑩\\
&+\frac{1}{2}\left\{\left(\frac{1}{3}\mathbf{u}^{dq+m}\bar{\mathbf{i}}^{dq+k}+\frac{1}{3}\mathbf{u}^{dq-m}\bar{\mathbf{i}}^{dq-k}+\frac{1}{2}\mathbf{u}^{dq0m}\bar{\mathbf{i}}^{dq0k}\right)e^{j((m-k)\omega t)}\right. && ⑪\\
&\qquad\left.+\left(\frac{1}{3}\bar{\mathbf{u}}^{dq+m}\mathbf{i}^{dq+k}+\frac{1}{3}\bar{\mathbf{u}}^{dq-m}\mathbf{i}^{dq-k}+\frac{1}{2}\bar{\mathbf{u}}^{dq0m}\mathbf{i}^{dq0k}\right)e^{-j((m-k)\omega t)}\right\} && ⑫
\end{aligned}
$$

The above multiplication yields in the following results, depending on $m > k$, $m < k$, and $m = k$. Define the complex space vectors with the nonitalic bold letters $\breve{\mathbf{p}}^{dq\pm n} = \breve{p}_{d\pm n} + j\cdot\breve{p}_{q\pm n}$, $\breve{\mathbf{p}}^{dq0n} = \breve{p}_{d0n} + j\cdot\breve{p}_{q0n}$, and the vectors with italic bold letters $\breve{\boldsymbol{p}}^{dq\pm n} = [\breve{p}_{d\pm n} \quad \breve{p}_{q\pm n}]^\top$, $\breve{\boldsymbol{p}}^{dq0n} = [\breve{p}_{d0n} \quad \breve{p}_{q0n}]^\top$.

1) Case $m > k$, from the phase shift in the terms in $\breve{p}_{ab}$, $\breve{p}_{bc}$ and $\breve{p}_{ca}$, it can be found that ① and ②, ③ and ④, ⑤ and ⑥ are respectively the complex conjugated terms for the $(m+k)$th positive-, negative-, and zero-sequence; ⑦ and

⑧, ⑨ and ⑩, ⑪ and ⑫ are respectively the complex conjugated terms for the $(m+k)$th positive-, negative-, and zero-sequence. The complex space vectors are in Eqs. (A.2.4)-(A.2.9).

$$\breve{\mathbf{p}}^{dq+(m+k)} = \frac{1}{2}\mathbf{u}^{dq0m}\mathbf{i}^{dq+k} + \frac{1}{\sqrt{6}}\mathbf{u}^{dq-m}\mathbf{i}^{dq-k} + \frac{1}{2}\mathbf{u}^{dq+m}\mathbf{i}^{dq0k} \tag{A.2.4}$$

$$\breve{\mathbf{p}}^{dq-(m+k)} = \frac{1}{\sqrt{6}}\mathbf{u}^{dq+m}\mathbf{i}^{dq+k} + \frac{1}{2}\mathbf{u}^{dq0m}\mathbf{i}^{dq-k} + \frac{1}{2}\mathbf{u}^{dq-m}\mathbf{i}^{dq0k} \tag{A.2.5}$$

$$\breve{\mathbf{p}}^{dq0(m+k)} = \frac{1}{3}\mathbf{u}^{dq-m}\mathbf{i}^{dq+k} + \frac{1}{3}\mathbf{u}^{dq+m}\mathbf{i}^{dq-k} + \frac{1}{2}\mathbf{u}^{dq0m}\mathbf{i}^{dq0k} \tag{A.2.6}$$

$$\breve{\mathbf{p}}^{dq+(m-k)} = \frac{1}{\sqrt{6}}\mathbf{u}^{dq-m}\bar{\mathbf{i}}^{dq+k} + \frac{1}{2}\mathbf{u}^{dq0m}\bar{\mathbf{i}}^{dq-k} + \frac{1}{2}\mathbf{u}^{dq+m}\bar{\mathbf{i}}^{dq0k} \tag{A.2.7}$$

$$\breve{\mathbf{p}}^{dq-(m-k)} = \frac{1}{2}\mathbf{u}^{dq0m}\bar{\mathbf{i}}^{dq+k} + \frac{1}{\sqrt{6}}\mathbf{u}^{dq+m}\bar{\mathbf{i}}^{dq-k} + \frac{1}{2}\mathbf{u}^{dq-m}\bar{\mathbf{i}}^{dq0k} \tag{A.2.8}$$

$$\breve{\mathbf{p}}^{dq0(m-k)} = \frac{1}{3}\mathbf{u}^{dq+m}\bar{\mathbf{i}}^{dq+k} + \frac{1}{3}\mathbf{u}^{dq-m}\bar{\mathbf{i}}^{dq-k} + \frac{1}{2}\mathbf{u}^{dq0m}\bar{\mathbf{i}}^{dq0k} \tag{A.2.9}$$

Separating the real and imaginary component of the complex space vectors in Eqs. (A.2.4)-(A.2.9), Eqs. (4.30) and (4.31) can be obtained by putting $\breve{\boldsymbol{p}}^{dq\pm(m+k)}$ and $\breve{\boldsymbol{p}}^{dq0(m+k)}$ into a vector $\breve{\boldsymbol{p}}^{dq(m+k)}$, $\breve{\boldsymbol{p}}^{dq\pm(m-k)}$ and $\breve{\boldsymbol{p}}^{dq0(m-k)}$ into a vector $\breve{\boldsymbol{p}}^{dq(m-k)}$. In Eqs. (4.30) and (4.31), $\boldsymbol{F}_1(\boldsymbol{u}^{dq\pm m})$, $\boldsymbol{F}_1(\boldsymbol{u}^{dq0m})$, $\boldsymbol{F}_2(\boldsymbol{u}^{dq\pm m})$ and $\boldsymbol{F}_2(\boldsymbol{u}^{dq0m})$ can be found in Eq. (A.2.16).

2) Case $m < k$, ① and ②, ③ and ④, ⑤ and ⑥ are respectively the complex conjugated terms for the $(m+k)$th positive-, negative-, and zero-sequence; ⑩ and ⑨, ⑧ and ⑦, ⑫ and ⑪ are respectively the complex conjugated terms for the $(m+k)$th positive-, negative-, and zero-sequence. The complex space vectors are given in Eqs. (A.2.10)-(A.2.12).

$$\breve{\mathbf{p}}^{dq+(k-m)} = \frac{1}{2}\bar{\mathbf{u}}^{dq0m}\mathbf{i}^{dq+k} + \frac{1}{\sqrt{6}}\bar{\mathbf{u}}^{dq+m}\mathbf{i}^{dq-k} + \frac{1}{2}\bar{\mathbf{u}}^{dq-m}\mathbf{i}^{dq0k} \tag{A.2.10}$$

$$\breve{\mathbf{p}}^{dq-(k-m)} = \frac{1}{\sqrt{6}}\bar{\mathbf{u}}^{dq-m}\mathbf{i}^{dq+k} + \frac{1}{2}\bar{\mathbf{u}}^{dq0m}\mathbf{i}^{dq-k} + \frac{1}{2}\bar{\mathbf{u}}^{dq+m}\mathbf{i}^{dq0k} \tag{A.2.11}$$

$$\breve{\mathbf{p}}^{dq0(k-m)} = \frac{1}{3}\bar{\mathbf{u}}^{dq+m}\mathbf{i}^{dq+k} + \frac{1}{3}\bar{\mathbf{u}}^{dq-m}\mathbf{i}^{dq-k} + \frac{1}{2}\bar{\mathbf{u}}^{dq0m}\mathbf{i}^{dq0k} \tag{A.2.12}$$

Separating the real and imaginary component, Eq. (4.33) can be obtained, in which $\breve{\boldsymbol{p}}^{dq\pm(k-m)}$ and $\breve{\boldsymbol{p}}^{dq0(k-m)}$ have been put into the vector $\breve{\boldsymbol{p}}^{dq(k-m)}$. In Eq. (4.33), $\boldsymbol{F}_3(\boldsymbol{u}^{dq\pm m})$ and $\boldsymbol{F}_3(\boldsymbol{u}^{dq0m})$ can be found in Eq. (A.2.16).

3) Case $m = k$, ① and ②, ③ and ④, ⑤ and ⑥ are respectively the complex conjugated terms for the $(m+k)$th positive-, negative-, and zero-sequence; ⑦ and

⑧ , ⑨ and ⑩ , ⑪ and ⑫ are the dc component with the following expression,

$$
\begin{aligned}
\breve{p}_{ab,dc} = {} & \sqrt{\frac{2}{3}}\,\mathrm{Re}\left\{\frac{1}{\sqrt{6}}\mathbf{u}^{dq-m}\bar{\mathbf{i}}^{dq+k} + \frac{1}{2}\mathbf{u}^{dq0m}\bar{\mathbf{i}}^{dq-k} + \frac{1}{2}\mathbf{u}^{dq+m}\bar{\mathbf{i}}^{dq0k}\right\} \\
& + \sqrt{\frac{2}{3}}\,\mathrm{Re}\left\{\frac{1}{2}\mathbf{u}^{dq0m}\bar{\mathbf{i}}^{dq+k} + \frac{1}{\sqrt{6}}\mathbf{u}^{dq+m}\bar{\mathbf{i}}^{dq-k} + \frac{1}{2}\mathbf{u}^{dq-m}\bar{\mathbf{i}}^{dq0k}\right\} \\
& + \mathrm{Re}\left\{\frac{1}{3}\mathbf{u}^{dq+m}\bar{\mathbf{i}}^{dq+k} + \frac{1}{3}\mathbf{u}^{dq-m}\bar{\mathbf{i}}^{dq-k} + \frac{1}{2}\mathbf{u}^{dq0m}\bar{\mathbf{i}}^{dq0k}\right\}
\end{aligned} \tag{A.2.13}
$$

$$
\begin{aligned}
\breve{p}_{bc,dc} = {} & \sqrt{\frac{2}{3}}\cos\left(\frac{2\pi}{3}\right)\mathrm{Re}\left\{\frac{1}{\sqrt{6}}\mathbf{u}^{dq-m}\bar{\mathbf{i}}^{dq+k} + \frac{1}{2}\mathbf{u}^{dq0m}\bar{\mathbf{i}}^{dq-k} + \frac{1}{2}\mathbf{u}^{dq+m}\bar{\mathbf{i}}^{dq0k}\right\} \\
& + \sqrt{\frac{2}{3}}\sin\left(\frac{2\pi}{3}\right)\mathrm{Im}\left\{\frac{1}{\sqrt{6}}\mathbf{u}^{dq-m}\bar{\mathbf{i}}^{dq+k} + \frac{1}{2}\mathbf{u}^{dq0m}\bar{\mathbf{i}}^{dq-k} + \frac{1}{2}\mathbf{u}^{dq+m}\bar{\mathbf{i}}^{dq0k}\right\} \\
& + \sqrt{\frac{2}{3}}\cos\left(\frac{2\pi}{3}\right)\mathrm{Re}\left\{\frac{1}{2}\mathbf{u}^{dq0m}\bar{\mathbf{i}}^{dq+k} + \frac{1}{\sqrt{6}}\mathbf{u}^{dq+m}\bar{\mathbf{i}}^{dq-k} + \frac{1}{2}\mathbf{u}^{dq-m}\bar{\mathbf{i}}^{dq0k}\right\} \\
& - \sqrt{\frac{2}{3}}\sin\left(\frac{2\pi}{3}\right)\mathrm{Im}\left\{\frac{1}{2}\mathbf{u}^{dq0m}\bar{\mathbf{i}}^{dq+k} + \frac{1}{\sqrt{6}}\mathbf{u}^{dq+m}\bar{\mathbf{i}}^{dq-k} + \frac{1}{2}\mathbf{u}^{dq-m}\bar{\mathbf{i}}^{dq0k}\right\} \\
& + \mathrm{Re}\left\{\frac{1}{3}\mathbf{u}^{dq+m}\bar{\mathbf{i}}^{dq+k} + \frac{1}{3}\mathbf{u}^{dq-m}\bar{\mathbf{i}}^{dq-k} + \frac{1}{2}\mathbf{u}^{dq0m}\bar{\mathbf{i}}^{dq0k}\right\}
\end{aligned} \tag{A.2.14}
$$

$$
\begin{aligned}
\breve{p}_{ca,dc} = {} & \sqrt{\frac{2}{3}}\cos\left(\frac{2\pi}{3}\right)\mathrm{Re}\left\{\frac{1}{\sqrt{6}}\mathbf{u}^{dq-m}\bar{\mathbf{i}}^{dq+k} + \frac{1}{2}\mathbf{u}^{dq0m}\bar{\mathbf{i}}^{dq-k} + \frac{1}{2}\mathbf{u}^{dq+m}\bar{\mathbf{i}}^{dq0k}\right\} \\
& - \sqrt{\frac{2}{3}}\sin\left(\frac{2\pi}{3}\right)\mathrm{Im}\left\{\frac{1}{\sqrt{6}}\mathbf{u}^{dq-m}\bar{\mathbf{i}}^{dq+k} + \frac{1}{2}\mathbf{u}^{dq0m}\bar{\mathbf{i}}^{dq-k} + \frac{1}{2}\mathbf{u}^{dq+m}\bar{\mathbf{i}}^{dq0k}\right\} \\
& + \sqrt{\frac{2}{3}}\cos\left(\frac{2\pi}{3}\right)\mathrm{Re}\left\{\frac{1}{2}\mathbf{u}^{dq0m}\bar{\mathbf{i}}^{dq+k} + \frac{1}{\sqrt{6}}\mathbf{u}^{dq+m}\bar{\mathbf{i}}^{dq-k} + \frac{1}{2}\mathbf{u}^{dq-m}\bar{\mathbf{i}}^{dq0k}\right\} \\
& + \sqrt{\frac{2}{3}}\sin\left(\frac{2\pi}{3}\right)\mathrm{Im}\left\{\frac{1}{2}\mathbf{u}^{dq0m}\bar{\mathbf{i}}^{dq+k} + \frac{1}{\sqrt{6}}\mathbf{u}^{dq+m}\bar{\mathbf{i}}^{dq-k} + \frac{1}{2}\mathbf{u}^{dq-m}\bar{\mathbf{i}}^{dq0k}\right\} \\
& + \mathrm{Re}\left\{\frac{1}{3}\mathbf{u}^{dq+m}\bar{\mathbf{i}}^{dq+k} + \frac{1}{3}\mathbf{u}^{dq-m}\bar{\mathbf{i}}^{dq-k} + \frac{1}{2}\mathbf{u}^{dq0m}\bar{\mathbf{i}}^{dq0k}\right\}
\end{aligned} \tag{A.2.15}
$$

which can be simplified to Eq. (4.35), collecting $\breve{p}_{ab,dc}$, $\breve{p}_{bc,dc}$ and $\breve{p}_{ca,dc}$ into a vector.

$$
\begin{aligned}
& \boldsymbol{F}_1(\boldsymbol{u}^{dq\lambda m}) = \begin{bmatrix} u_{d\lambda m} & -u_{q\lambda m} \\ u_{q\lambda m} & u_{d\lambda m} \end{bmatrix}, \quad \boldsymbol{F}_2(\boldsymbol{u}^{dq\lambda m}) = \begin{bmatrix} u_{d\lambda m} & u_{q\lambda m} \\ u_{q\lambda m} & -u_{d\lambda m} \end{bmatrix} \\
& \boldsymbol{F}_3(\boldsymbol{u}^{dq\lambda m}) = \begin{bmatrix} u_{d\lambda m} & u_{q\lambda m} \\ -u_{q\lambda m} & u_{d\lambda m} \end{bmatrix}, \quad \text{with} \quad \lambda \in \{+, -, 0\}
\end{aligned} \tag{A.2.16}
$$

A.3. Harmonic Sequence Coupling

Table A.1.: Harmonic sequences in powers brought by the m-th branch voltage $\boldsymbol{u}^{abcm}$ and the k-th branch current $\boldsymbol{i}^{abck}$, where $m > k$

$\breve{\boldsymbol{p}}^{abc}$ \ $\boldsymbol{i}^{abck}$, $\boldsymbol{u}^{abcm}$	$\boldsymbol{i}^{abc+k}$	$\boldsymbol{i}^{abc-k}$	$\boldsymbol{i}^{abc0k}$
$\boldsymbol{u}^{abc+m}$	$0(m-k)$, $-(m+k)$	$-(m-k)$, $0(m+k)$	$+(m-k)$, $+(m+k)$
$\boldsymbol{u}^{abc-m}$	$+(m-k)$, $0(m+k)$	$0(m-k)$, $+(m+k)$	$-(m-k)$, $-(m+k)$
$\boldsymbol{u}^{abc0m}$	$-(m-k)$, $+(m+k)$	$+(m-k)$, $-(m+k)$	$0(m-k)$, $0(m+k)$

Table A.2.: Harmonic sequences in powers brought by the m-th branch voltage $\boldsymbol{u}^{abcm}$ and the k-th branch current $\boldsymbol{i}^{abck}$, where $m < k$

$\breve{\boldsymbol{p}}^{abc}$ \ $\boldsymbol{i}^{abck}$, $\boldsymbol{u}^{abcm}$	$\boldsymbol{i}^{abc+k}$	$\boldsymbol{i}^{abc-k}$	$\boldsymbol{i}^{abc0k}$
$\boldsymbol{u}^{abc+m}$	$0(k-m)$, $-(m+k)$	$+(k-m)$, $0(m+k)$	$-(k-m)$, $+(m+k)$
$\boldsymbol{u}^{abc-m}$	$-(k-m)$, $0(m+k)$	$0(k-m)$, $+(m+k)$	$+(k-m)$, $-(m+k)$
$\boldsymbol{u}^{abc0m}$	$+(k-m)$, $+(m+k)$	$-(k-m)$, $-(m+k)$	$0(k-m)$, $0(m+k)$

Table A.3.: Harmonic sequences in powers brought by the m-th branch voltage $\boldsymbol{u}^{abcm}$ and the k-th branch current $\boldsymbol{i}^{abck}$, where $m = k$

$\breve{\boldsymbol{p}}^{abc}$ \ $\boldsymbol{i}^{abck}$, $\boldsymbol{u}^{abcm}$	$\boldsymbol{i}^{abc+k}$	$\boldsymbol{i}^{abc-k}$	$\boldsymbol{i}^{abc0k}$
$\boldsymbol{u}^{abc+m}$	*dc*, $-(m+k)$	*dc unbal*, $0(m+k)$	*dc unbal*, $+(m+k)$
$\boldsymbol{u}^{abc-m}$	*dc unbal*, $0(m+k)$	*dc*, $+(m+k)$	*dc unbal*, $-(m+k)$
$\boldsymbol{u}^{abc0m}$	*dc unbal*, $+(m+k)$	*dc unbal*, $-(m+k)$	*dc*, $0(m+k)$
where *dc unbal* means that the *dc* quantity in each branch is different, but their sum is 0; *dc* stands for a same *dc* quantity in three branches			

Bibliography

[ABH07] Asiminoael, L. ; Blaabjerg, F. ; Hansen, S.: Detection is key - Harmonic detection methods for active power filter applications. In: *IEEE Industry Applications Magazine* (2007)

[AIY07] Akagi, H. ; Inoue, S. ; Yoshii, T.: Control and Performance of a Transformerless Cascade PWM STATCOM With Star Configuration. In: *IEEE Transactions on Industry Applications* (2007)

[AJKM16] Antoniewicz, K. ; Jasinski, M. ; Kazmierkowski, M. P. ; Malinowski, M.: Model Predictive Control for Three-Level Four-Leg Flying Capacitor Converter Operating as Shunt Active Power Filter. In: *IEEE Transactions on Industrial Electronics* (2016)

[Ans09] Anstreicher, Kurt M.: Semidefinite programming versus the reformulation-linearization technique for nonconvex quadratically constrained quadratic programming. In: *Journal of Global Optimization* (2009)

[AR99] Alves, M. F. ; Ribeiro, T. N.: Voltage sag: an overview of IEC and IEEE standards and application criteria. In: *1999 IEEE Transmission and Distribution Conference (Cat. No. 99CH36333)* (1999)

[AROARd+15] A. Ribeiro, R. L. ; O. A. Rocha, T. d. ; de Sousa, R. M. ; dos Santos, E. C. ; Lima, A. M. N.: A Robust DC-Link Voltage Control Strategy to Enhance the Performance of Shunt Active Power Filters Without Harmonic Detection Schemes. In: *IEEE Transactions on Industrial Electronics* (2015)

[Bak80] BAKER, R. H.: Switching circuit / U.S. Patent 4 210 826. July 1980. – Forschungsbericht

[BB17] BEHROUZIAN, E. ; BONGIORNO, M.: Investigation of Negative-Sequence Injection Capability of Cascaded H-Bridge Converters in Star and Delta Configuration. In: *IEEE Transactions on Power Electronics* (2017)

[BBM17] BORRELLI, Francesco ; BEMPORAD, Alberto ; MORARI, Manfred: *Predictive Control for Linear and Hybrid Systems*. Cambridge University Press, 2017

[BDSD18] BERGNA-DIAZ, G. ; SUUL, J. A. ; D'ARCO, S.: Energy-Based State-Space Representation of Modular Multilevel Converters with a Constant Equilibrium Point in Steady-State Operation. In: *IEEE Transactions on Power Electronics* (2018)

[Bie13] BIELA, A. Hillers; J.: Optimal design of the modular multilevel converter for an energy storage system based on split batteries. In: *2013 15th European Conference on Power Electronics and Applications (EPE)*, 2013

[BJH+15] BAOAN, Wang ; JIAO, Shang ; HAO, Chen ; LI, L. ; NINGYI, Dai: Reduction of converter rating for a delta-connected STATCOM by optimizing phase harmonic references. In: *TENCON 2015 - 2015 IEEE Region 10 Conference*, 2015

[BM13] BARUSCHKA, L. ; MERTENS, A.: A New Three-Phase AC/AC Modular Multilevel Converter With Six Branches in Hexagonal Configuration. In: *IEEE Transactions on Industry Applications* (2013)

[BMLB15a] BAKHSHIZADEH, M. K. ; MA, K. ; LOH, P. C. ; BLAABJERG, F.: Improvement of device current ratings in Modular Multilevel Converter by utilizing circulating current. In: *2015 IEEE Energy Conversion Congress and Exposition (ECCE)*, 2015

[BMLB15b] BAKHSHIZADEH, M. K. ; MA, K. ; LOH, P. C. ; BLAABJERG, F.:

Indirect thermal control for improved reliability of Modular Multilevel Converter by utilizing circulating current. In: *2015 IEEE Applied Power Electronics Conference and Exposition (APEC)*, 2015

[Bon85] BONNARD, G.: The problems posed by electrical power supply to industrial installations. In: *IEE Proceedings B - Electric Power Applications* (1985)

[BST11] BAO, Xiaowei ; SAHINIDIS, Nikolaos V. ; TAWARMALANI, Mohit: Semidefinite relaxations for quadratically constrained quadratic programming: A review and comparisons. In: *Mathematical Programming* (2011)

[BTLT06] BLAABJERG, F. ; TEODORESCU, R. ; LISERRE, M. ; TIMBUS, A. V.: Overview of Control and Grid Synchronization for Distributed Power Generation Systems. In: *IEEE Transactions on Industrial Electronics* (2006)

[CHX+04] CHANG, G. ; HATZIADONIU, C. ; XU, W. ; RIBEIRO, P. ; BURCH, R. ; GRADY, W. M. ; HALPIN, M. ; LIU, Y. ; RANADE, S. ; RUTHMAN, D. ; WATSON, N. ; ORTMEYER, T. ; WIKSTON, J. ; MEDINA, A. ; TESTA, A. ; GARDINIER, R. ; DINAVAHI, V. ; ACRAM, F. ; LEHN, P.: Modeling devices with nonlinear Voltage-current Characteristics for harmonic studies. In: *IEEE Transactions on Power Delivery* (2004)

[CKST12] CRACIUN, B. I. ; KEREKES, T. ; SERA, D. ; TEODORESCU, R.: Overview of recent Grid Codes for PV power integration. In: *2012 13th International Conference on Optimization of Electrical and Electronic Equipment (OPTIM)*, 2012

[CLC12] CHEN, Z. ; LUO, Y. ; CHEN, M.: Control and Performance of a Cascaded Shunt Active Power Filter for Aircraft Electric Power System. In: *IEEE Transactions on Industrial Electronics* (2012)

[CLL00] CHENG, Chuan-Ping ; LIU, Chih-Wen ; LIU, Chun-Chang: Unit commitment by Lagrangian relaxation and genetic algorithms. In:

Power Systems, IEEE Transactions on 15 (2000), S. 707–714

[CWL+15] Chen, H. C. ; Wu, P. H. ; Lee, C. T. ; Wang, C. W. ; Yang, C. H. ; Cheng, P. T.: Zero-Sequence Voltage Injection for DC Capacitor Voltage Balancing Control of the Star-Connected Cascaded H-Bridge PWM Converter Under Unbalanced Grid. In: *IEEE Transactions on Industry Applications* (2015)

[DCE+17] Diaz, M. ; Cardenas, R. ; Espinoza, M. ; Rojas, F. ; Mora, A. ; Clare, J. C. ; Wheeler, P.: Control of Wind Energy Conversion Systems Based on the Modular Multilevel Matrix Converter. In: *IEEE Transactions on Industrial Electronics* (2017)

[DGMD12] Dinh, Quoc T. ; Gumussoy, S. ; Michiels, W. ; Diehl, M.: Combining Convex - Concave Decompositions and Linearization Approaches for Solving BMIs, With Application to Static Output Feedback. In: *Automatic Control, IEEE Transactions on* 57 (2012), S. 1377–1390

[DVLP+13] Debrouwere, F. ; Van Loock, W. ; Pipeleers, G. ; Dinh, Q.T. ; Diehl, M. ; De Schutter, J. ; Swevers, J.: Time-Optimal Path Following for Robots With Convex-Concave Constraints Using Sequential Convex Programming. In: *Robotics, IEEE Transactions on* 29 (2013), S. 1485–1495

[DXM16] Dian, R. ; Xu, W. ; Mu, C.: Improved Negative Sequence Current Detection and Control Strategy for H-Bridge Three-Level Active Power Filter. In: *IEEE Transactions on Applied Superconductivity* (2016)

[EAN01] Erickson, R. W. ; Al-Naseem, O. A.: A new family of matrix converters. In: *Industrial Electronics Society, 2001. IECON '01. The 27th Annual Conference of the IEEE*, 2001

[EuS] *Standard EN 50160 - Voltage characteristics of electricity supplied by public distribution systems*

[EVMMRZ07] Escobar, G. ; Valdez, A. A. ; Martinez-Montejano, M. F. ;

RODRIGUEZ-ZERMENO, V. M.: A Model-Based Controller for the Cascade Multilevel Converter Used as a Shunt Active Filter. In: *2007 IEEE Industry Applications Annual Meeting*, 2007

[FAD09] FLOURENTZOU, N. ; AGELIDIS, V. G. ; DEMETRIADES, G. D.: VSC-Based HVDC Power Transmission Systems: An Overview. In: *IEEE Transactions on Power Electronics* (2009)

[FHA15] FARIVAR, G. ; HREDZAK, B. ; AGELIDIS, V. G.: Reduced-Capacitance Thin-Film H-Bridge Multilevel STATCOM Control Utilizing an Analytic Filtering Scheme. In: *IEEE Transactions on Industrial Electronics* (2015)

[GDM15] GEYER, T. ; DARIVIANAKIS, G. ; MERWE, W. van d.: Model predictive control of a STATCOM based on a modular multilevel converter in delta configuration. In: *2015 17th European Conference on Power Electronics and Applications (EPE'15 ECCE-Europe)*, 2015

[Gen06] GENE F. FRANKLIN, J. DAVID POWELL AND MICHAEL WORKMAN: *Digital Control of Dynamic Systems*. Ellis-Kagle Press, since 2006

[HAA07] HIROFUMI AKAGI, Edson Hirokazu W. ; AREDES, Mauricio: *Instantaneous Power Theory and Applications to Power Conditioning*. Wiley-IEEE Press, April 2007

[HDS98] HANIF D. SHERALI, Warren P. A.: *A Reformulation-Linearization Technique for Solving Discrete and Continuous Nonconvex Problems*. Springer US, 1998

[HKE14] HAWKE, J. T. ; KRISHNAMOORTHY, H. S. ; ENJETI, P. N.: A Family of New Multiport Power-Sharing Converter Topologies for Large Grid-Connected Fuel Cells. In: *IEEE Journal of Emerging and Selected Topics in Power Electronics* (2014)

[HKHC10] HAN, Sang wook ; KIM, Hoon ; HAN, Youngnam ; CIOFFI, J.M.: Efficient power allocation schemes for nonconvex sum-rate maximization on gaussian cognitive MAC. In: *Communications, IEEE Transactions on* 58 (2010), S. 753–757

[HMA12] Hagiwara, M. ; Maeda, R. ; Akagi, H.: Negative-Sequence Reactive-Power Control by a PWM STATCOM Based on a Modular Multilevel Cascade Converter (MMCC-SDBC). In: *IEEE Transactions on Industry Applications* (2012)

[IANN12] Ilves, K. ; Antonopoulos, A. ; Norrga, S. ; Nee, H. P.: Steady-State Analysis of Interaction Between Harmonic Components of Arm and Line Quantities of Modular Multilevel Converter. In: *IEEE Transactions on Power Electronics* (2012)

[IEE94] IEEE Recommended Practice for Electric Power Distribution for Industrial Plants. In: *IEEE Std 141-1993* (1994)

[IEE09] IEEE Recommended Practice for Monitoring Electric Power Quality. In: *IEEE Std 1159-2009 (Revision of IEEE Std 1159-1995)* (2009)

[IEE14] IEEE: IEEE Recommended Practice and Requirements for Harmonic Control in Electric Power Systems. In: *IEEE Std 519-2014 (Revision of IEEE Std 519-1992)* (2014)

[JJ15] Jovcic, D. ; Jamshidifar, A. A.: Phasor Model of Modular Multilevel Converter With Circulating Current Suppression Control. In: *IEEE Transactions on Power Delivery* (2015)

[JJ16] Jamshidifar, A. ; Jovcic, D.: Small-Signal Dynamic DQ Model of Modular Multilevel Converter for System Studies. In: *IEEE Transactions on Power Delivery* (2016)

[KBM15] Karwatzki, D. ; Baruschka, L. ; Mertens, A.: Survey on the Hexverter topology : A modular multilevel AC/AC converter. In: *2015 9th International Conference on Power Electronics and ECCE Asia (ICPE-ECCE Asia)*, 2015

[KHA14] Kawamura, W. ; Hagiwara, M. ; Akagi, H.: Control and

Experiment of a Modular Multilevel Cascade Converter Based on Triple-Star Bridge Cells. In: *IEEE Transactions on Industry Applications* (2014)

[KHS15] KANDPAL, M. ; HUSSAIN, I. ; SINGH, B.: Grid integration of solar PV generating system using three-level voltage source converter. In: *2015 Annual IEEE India Conference (INDICON)*, 2015

[KKGB15] KOLB, J. ; KAMMERER, F. ; GOMMERINGER, M. ; BRAUN, M.: Cascaded Control System of the Modular Multilevel Converter for Feeding Variable-Speed Drives. In: *IEEE Transactions on Power Electronics* (2015)

[KKZ13] KANJIYA, P. ; KHADKIKAR, V. ; ZEINELDIN, H.H.: A Noniterative Optimized Algorithm for Shunt Active Power Filter Under Distorted and Unbalanced Supply Voltages. In: *Industrial Electronics, IEEE Transactions on* 60 (2013), S. 5376–5390

[Kol14] KOLB, Johannes: *Optimale Betriebsfuehrung des Modularen Multilevel-Umrichters als Antriebsumrichter fuer Drehstrommaschinen.* KIT Scientific Publishing, Karlsruhe, 2014

[KTHC16] KRASTEV, I. ; TRICOLI, P. ; HILLMANSEN, S. ; CHEN, M.: Future of Electric Railways: Advanced Electrification Systems with Static Converters for ac Railways. In: *IEEE Electrification Magazine* (2016)

[KV17] KUMAR, A. ; VERMA, V.: Analysis and control of improved power quality single-phase split voltage cascaded converter feeding three-phase OEIM drive. In: *IET Power Electronics* (2017)

[KWB+16] KWON, J. B. ; WANG, X. ; BLAABJERG, F. ; BAK, C. L. ; WOOD, A. R. ; WATSON, N. R.: Harmonic Instability Analysis of a Single-Phase Grid-Connected Converter Using a Harmonic State-Space Modeling Method. In: *IEEE Transactions on Industry Applications* (2016)

[KWB+17] KWON, J. ; WANG, X. ; BLAABJERG, F. ; BAK, C. L. ; SU-

LAREA, V. ; BUSCA, C.: Harmonic Interaction Analysis in a Grid-Connected Converter Using Harmonic State-Space (HSS) Modeling. In: *IEEE Transactions on Power Electronics* (2017)

[KWS10] KORN, A. J. ; WINKELNKEMPER, M. ; STEIMER, P.: Low output frequency operation of the Modular Multi-Level Converter. In: *2010 IEEE Energy Conversion Congress and Exposition*, 2010

[LCM16] LYU, J. ; CAI, X. ; MOLINAS, M.: Frequency Domain Stability Analysis of MMC-Based HVdc for Wind Farm Integration. In: *IEEE Journal of Emerging and Selected Topics in Power Electronics* (2016)

[LMS+10] LUO, Zhi-Quan ; MA, Wing-Kin ; SO, AM.-C. ; YE, Yinyu ; ZHANG, Shuzhong: Semidefinite Relaxation of Quadratic Optimization Problems. In: *Signal Processing Magazine, IEEE* 27 (2010), S. 20–34

[Lor01] LORENZ, R. D.: Robotics and automation applications of drives and converters. In: *Proceedings of the IEEE* (2001)

[LQTH16] LI, S. ; QI, W. ; TAN, S. ; HUI, S. Y.: Integration of an Active Filter and a Single-Phase AC/DC Converter With Reduced Capacitance Requirement and Component Count. In: *IEEE Transactions on Power Electronics* (2016)

[LSN+14] LAWDER, M. T. ; SUTHAR, B. ; NORTHROP, P. W. C. ; DE, S. ; HOFF, C. M. ; LEITERMANN, O. ; CROW, M. L. ; SANTHANAGOPALAN, S. ; SUBRAMANIAN, V. R.: Battery Energy Storage System (BESS) and Battery Management System (BMS) for Grid-Scale Applications. In: *Proceedings of the IEEE* (2014)

[LWY+17] LU, D. ; WANG, J. ; YAO, J. ; WANG, S. ; ZHU, J. ; HU, H. ; ZHANG, L.: Clustered Voltage Balancing Mechanism and Its Control Strategy for Star-Connected Cascaded H-Bridge STATCOM. In: *IEEE Transactions on Industrial Electronics* (2017)

[LYX+15] LI, B. ; YANG, R. ; XU, D. ; WANG, G. ; WANG, W. ; XU,

D.: Analysis of the Phase-Shifted Carrier Modulation for Modular Multilevel Converters. In: *IEEE Transactions on Power Electronics* (2015)

[LZX+16] LI, B. ; ZHOU, S. ; XU, D. ; YANG, R. ; XU, D. ; BUCCELLA, C. ; CECATI, C.: An Improved Circulating Current Injection Method for Modular Multilevel Converters in Variable-Speed Drives. In: *IEEE Transactions on Industrial Electronics* (2016)

[MCG07] MONTERO, M.IM. ; CADAVAL, E.R. ; GONZALEZ, F.B.: Comparison of Control Strategies for Shunt Active Power Filters in Three-Phase Four-Wire Systems. In: *Power Electronics, IEEE Transactions on* 22 (2007), S. 229–236

[MFYI17] MIURA, Y. ; FUJIKAWA, T. ; YOSHIDA, T. ; ISE, T.: Control scheme of the modular multilevel matrix converter using space vector modulation for wide frequency range operation. In: *2017 IEEE Energy Conversion Congress and Exposition (ECCE)*, 2017

[MH09] MCGRATH, Brendan P. ; HOLMES, Donald G.: A General Analytical Method for Calculating Inverter DC-Link Current Harmonics. In: *IEEE Transactions on Industrial Applications* (2009)

[MMT+17] MADHUSOODHANAN, S. ; MAINALI, K. ; TRIPATHI, A. ; PATEL, D. ; KADAVELUGU, A. ; BHATTACHARYA, S. ; HATUA, K.: Harmonic Analysis and Controller Design of 15 kV SiC IGBT-Based Medium-Voltage Grid-Connected Three-Phase Three-Level NPC Converter. In: *IEEE Transactions on Power Electronics* (2017)

[MnCGPA16] MORALES, J. L. M. ; ÁNGELES, M. H. ; CAMPOS-GAONA, D. ; PEÑA-ALZOLA, R.: Control design of a neutral point clamped converter based active power filter for the selective harmonic compensation. In: *2016 IEEE PES Transmission Distribution Conference and Exposition-Latin America (PES T D-LA)*, 2016

[MTMT10] MUYEEN, S. M. ; TAKAHASHI, R. ; MURATA, T. ; TAMURA, J.: A Variable Speed Wind Turbine Control Strategy to Meet Wind

Farm Grid Code Requirements. In: *IEEE Transactions on Power Systems* (2010)

[MUR03] MOHAN, Ned ; UNDELAND, Tore M. ; ROBBINS, William P.: *Power Electronics. CONVERTERS, APPLICATIONS AND DESIGN.* John Wiley and Sons, Inc, 2003

[NFCP+17] NAZARETH FERREIRA, V. de ; CUPERTINO, A. F. ; PEREIRA, H. A. ; ROCHA, A. V. ; SELEME, S. I. ; JESUS CARDOSO FILHO, B. de: Design of high-reliable converters for medium-voltage rolling mills systems. In: *2017 IEEE Industry Applications Society Annual Meeting*, 2017

[NL07] NINAD, N. A. ; LOPES, L. A. C.: Operation of Single-phase Grid-Connected Inverters with Large DC Bus Voltage Ripple. In: *2007 IEEE Canada Electrical Power Conference*, 2007

[NSBP17] NARULA, S. ; SINGH, B. ; BHUVANESWARI, G. ; PANDEY, R.: Improved Power Quality Bridgeless Converter-Based SMPS for Arc Welding. In: *IEEE Transactions on Industrial Electronics* (2017)

[PSMD14] PORRU, M. ; SERPI, A. ; MARONGIU, I. ; DAMIANO, A.: A novel DC-link voltage and current control algorithm for Neutral-Point-Clamped converters. In: *IECON 2014 - 40th Annual Conference of the IEEE Industrial Electronics Society*, 2014

[QZLY11] QIAN, H. ; ZHANG, J. ; LAI, J. S. ; YU, W.: A high-efficiency grid-tie battery energy storage system. In: *IEEE Transactions on Power Electronics* (2011)

[RFK+09] RODRIGUEZ, J. ; FRANQUELO, L.G. ; KOURO, S. ; LEON, J.I. ; PORTILLO, R.C. ; PRATS, M.A.M. ; PEREZ, M.A.: Multilevel Converters: An Enabling Technology for High-Power Applications. In: *Proceedings of the IEEE* 97 (2009), S. 1786–1817

[RKM18] RAOUFAT, M. E. ; KHAYATIAN, A. ; MOJALLAL, A.: Performance Recovery of Voltage Source Converters With Application to Grid-Connected Fuel Cell DGs. In: *IEEE Transactions on Smart Grid*

(2018)

[RMB+15] Rejas, M. ; Mathe, L. ; Burlacu, P. D. ; Pereira, H. ; Sangwongwanich, A. ; Bongiorno, M. ; Teodorescu, R.: Performance comparison of phase shifted PWM and sorting method for modular multilevel converters. In: *2015 17th European Conference on Power Electronics and Applications (EPE'15 ECCE-Europe)*, 2015

[RMG08] Rao, U.K. ; Mishra, M.K. ; Ghosh, A: Control Strategies for Load Compensation Using Instantaneous Symmetrical Component Theory Under Different Supply Voltages. In: *Power Delivery, IEEE Transactions on* 23 (2008), S. 2310–2317

[RTGG01] Rafiei, S.M.-R. ; Toliyat, H.A ; Ghazi, R. ; Gopalarathnam, T.: An optimal and flexible control strategy for active filtering and power factor correction under non-sinusoidal line voltages. In: *Power Delivery, IEEE Transactions on* 16 (2001), S. 297–305

[SA15] Shukla A., Nami A. ; Shahnia F., Ghosh A. Rajakaruna S. S. Rajakaruna S. (Hrsg.): *Multilevel Converter Topologies for STATCOMs.* Springer, Singapore, 2015

[SAT17] Sochor, P. ; Akagi, H. ; Tan, N. M. L.: Low-voltage-ride-through control of a modular multilevel SDBC inverter for utility-scale photovoltaic systems. In: *2017 IEEE Energy Conversion Congress and Exposition (ECCE)*, 2017

[SB99] Sergio Bittanti, Patrizio C. ; Electrical, Encyclopedia of (Hrsg.) ; Engineering, Electronics (Hrsg.): *Periodic Control Systems.* Wiley, 1999

[Sch99] Schauder, C.: STATCOM for compensation of large electric arc furnace installations. In: *1999 IEEE Power Engineering Society Summer Meeting. Conference Proceedings (Cat. No.99CH36364)*, 1999

[SG02] Soto, D. ; Green, T. C.: A comparison of high-power converter topologies for the implementation of FACTS controllers. In: *IEEE Transactions on Industrial Electronics* (2002)

[Sin09] SINGH, Sri N.: Distributed Generation in Power Systems: An Overview and Key Issues. In: *24th Indian Engineering Congress*, 2009

[SL14] SUBROTO, Ramadhani K. ; LIAN, Kuo L.: Modeling of a Multilevel Voltage Source Converter Using the Fast Time-Domain Method. In: *IEEE Journal of Emerging and Selected Topics in Power Electronics*, 2014

[SLZ+16] SHU, Z. ; LIU, M. ; ZHAO, L. ; SONG, S. ; ZHOU, Q. ; HE, X.: Predictive Harmonic Control and Its Optimal Digital Implementation for MMC-Based Active Power Filter. In: *IEEE Transactions on Industrial Electronics* (2016)

[SMSG+10] SONG-MANGUELLE, J. ; SCHRODER, S. ; GEYER, T. ; EKEMB, G. ; NYOBE-YOME, J. M.: Prediction of Mechanical Shaft Failures Due to Pulsating Torques of Variable-Frequency Drives. In: *IEEE Transactions on Industry Applications* (2010)

[SP04] SOTO, D. ; PENA, R.: Nonlinear control strategies for cascaded multilevel STATCOMs. In: *IEEE Transactions on Power Delivery* (2004)

[TDMMN15] TRINTIS, I. ; DOUGLASS, P. ; MAHESHWARI, R. ; MUNK-NIELSEN, S.: SiC heat pump converters with support for voltage unbalance in distribution grids. In: *2015 17th European Conference on Power Electronics and Applications (EPE'15 ECCE-Europe)*, 2015

[TDQ11] TRAN DINH QUOC, Moritz D.: Sequential Convex Programming Methods for Solving Nonlinear Optimization Problems with DC Constraints. Available: http://arxiv.org/abs/1107.5841, 2011

[TRP+15] THANTIRIGE, K. ; RATHORE, A. K. ; PANDA, S. K. ; JAYASIGNHE, G. ; ZAGRODNIK, M. A. ; GUPTA, A. K.: Medium voltage multilevel converters for ship electric propulsion drives. In: *2015 Inter-*

national Conference on Electrical Systems for Aircraft, Railway, Ship Propulsion and Road Vehicles (ESARS), 2015

[UMG09] Uyyuru, K.R. ; Mishra, M.K. ; Ghosh, A: An Optimization-Based Algorithm for Shunt Active Filter Under Distorted Supply Voltages. In: *Power Electronics, IEEE Transactions on* 24 (2009), S. 1223–1232

[VB96] Vandenberghe, Lieven ; Boyd, Stephen: Semidefinite Programming. In: *SIAM Review* 38 (1996), S. 49–95

[VDB+18] Valavi, M. ; Devillers, E. ; Besnerais, J. L. ; Nysveen, A. ; Nilsen, R.: Influence of Converter Topology and Carrier Frequency on Airgap Field Harmonics, Magnetic Forces, and Vibrations in Converter-Fed Hydropower Generator. In: *IEEE Transactions on Industry Applications* (2018)

[VEToMm06] Valdez, A. A. ; Escobar, G. ; Torres-olguin, R. E. ; Martinez-montejano, M. F.: A model-based controller for a three-phase four-wire shunt active filter with compensation of the neutral line current. In: *2006 IEEE International Power Electronics Congress*, 2006

[Wak10] Wakileh, George J.: *Power Systems Harmonics: Fundamentals, Analysis and Filter Design.* Springer Berlin Heidelberg, 2010

[WB14] Wang, H. ; Blaabjerg, F.: Reliability of Capacitors for DC-Link Applications in Power Electronic Converters - An Overview. In: *IEEE Transactions on Industry Applications* (2014)

[WCC17] Wu, P. H. ; Chen, Y. T. ; Cheng, P. T.: The Delta-Connected Cascaded H-Bridge Converter Application in Distributed Energy Resources and Fault Ride Through Capability Analysis. In: *IEEE Transactions on Industry Applications* (2017)

[WCCC17] Wu, P. H. ; Chen, H. C. ; Chang, Y. T. ; Cheng, P. T.: Delta-Connected Cascaded H-Bridge Converter Application in Unbalanced Load Compensation. In: *IEEE Transactions on Industry*

Applications (2017)

[WCT+16] WANG, Y. ; CHAI, Y. ; TANG, J. ; YUAN, X. ; XIA, C.: DC voltage control strategy of chain star STATCOM with second-order harmonic suppression. In: *IET Power Electronics* (2016)

[WHK05] WOOK HYUN KWON, Soo Hee H.: *Optimal Controls on Finite and Infinite Horizons: A Review. In: Receding Horizon Control.* Springer, London, 2005

[WHO14] WANG, C. ; HAO, Q. ; OOI, B.: Reduction of low-frequency harmonics in modular multilevel converters (MMCs) by harmonic function analysis. In: *IET Generation, Transmission Distribution* (2014)

[WKS10] WINKELNKEMPER, M. ; KORN, A. ; STEIMER, P.: A modular direct converter for transformerless rail interties. In: *2010 IEEE International Symposium on Industrial Electronics*, 2010

[WL15] WANG, H. ; LIU, S.: Adaptive Kalman filter for harmonic detection in active power filter application. In: *2015 IEEE Electrical Power and Energy Conference (EPEC)*, 2015

[WL16] WANG, H. ; LIU, S.: An optimal strategy with convex-concave constraints for power factor correction and harmonic compensation under different voltages. In: *2016 IEEE International Conference on Industrial Technology (ICIT)*, 2016

[WL17] WANG, H. ; LIU, S.: Capacitor voltage regulation of modular multilevel cascaded converter (MMCC-SDBC) as shunt active power filter under different PCC voltages. In: *IECON 2017 - 43rd Annual Conference of the IEEE Industrial Electronics Society*, 2017

[WL19a] WANG, Hengyi ; LIU, Steven: Harmonic Interaction Analysis of Delta-connected Cascaded H-bridge-based Shunt Active Power Filter. In: *IEEE Journal of Emerging and Selected Topics in Power Electronics* (July 2019)

[WL19b] WANG, Hengyi ; LIU, Steven: An optimal operation strategy for an active power filter using cascaded Hbridges in delta-connection. In: *Electric Power Systems Research* (October 2019)

[WLZX13] WANG, K. ; LI, Y. ; ZHENG, Z. ; XU, L.: Voltage Balancing and Fluctuation-Suppression Methods of Floating Capacitors in a New Modular Multilevel Converter. In: *IEEE Transactions on Industrial Electronics* (2013)

[WSDC15] WANG, B. ; SHANG, J. ; DAI, N. ; CHEN, H.: Harmonic reference currents balancing method for delta-connected static synchronous compensator. In: *Electronics Letters* (2015)

[WTWL14] WANG, H. ; TONG, J. ; WAN, Y. ; LIU, S.: Integrated current-energy modeling and nonlinear feedback control of modular multilevel STATCOM. In: *IECON 2014 - 40th Annual Conference of the IEEE Industrial Electronics Society*, 2014

[Xu00] XU, Wilsun: Comparisons and comments on harmonic standards IEC 1000-3-6 and IEEE Std. 519. In: *Ninth International Conference on Harmonics and Quality of Power. Proceedings (Cat. No.00EX441)*, 2000

[XYS07] XU, L. ; YAO, L. ; SASSE, C.: Grid Integration of Large DFIG-Based Wind Farms Using VSC Transmission. In: *IEEE Transactions on Power Systems* (2007)

[YKT+17] YU, Y. ; KONSTANTINOU, G. ; TOWNSEND, C. D. ; AGUILERA, R. P. ; AGELIDIS, V. G.: Delta-Connected Cascaded H-Bridge Multilevel Converters for Large-Scale Photovoltaic Grid Integration. In: *IEEE Transactions on Industrial Electronics* (2017)

[YSL+17] YANG, Z. ; SUN, J. ; LI, S. ; HUANG, M. ; ZHA, X. ; TANG, Y.: An Adaptive Carrier Frequency Optimization Method for Harmonic Energy Unbalance Minimization in a Cascaded H-Bridge-based Active Power Filter. In: *IEEE Transactions on Power Electronics* (2017)

[YWT+18] YANG, S. ; WANG, P. ; TANG, Y. ; ZAGRODNIK, M. ; HU, X. ; TSENG, K. J.: Circulating Current Suppression in Modular Multilevel Converters With Even-Harmonic Repetitive Control. In: *IEEE Transactions on Industry Applications* (2018)

Hengyi Wang

Erwin-Schrödinger-Str. 12 • 67663 Kaiserslautern • Email:hwang@eit.uni-kl.de

Lebenslauf

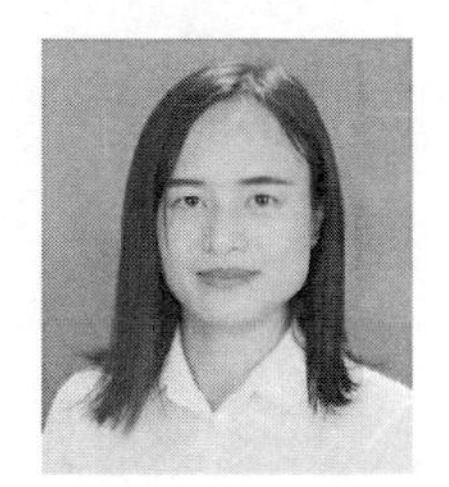

Persönliche Daten

Name:	Hengyi Wang
Geschlecht:	weiblich
Geburtsdaten:	02.08.1986
Nationalität:	V.R. China

Ausbildung

04.2012 – 08.2019	Promotion an der TU Kaiserslautern, Kaiserslautern, Deutschland Fachbereich: Elektrotechnik und Informationstechnik Finanzielle Unterstützung durch das DAAD Abschlussstipendium von 11.2017 bis 12.2017. Finanzielle Unterstützung vom Doktorvater von 04.2016 bis 10.2017. Finanzielle Unterstützung durch das China Scholarship Council (CSC) Stipendium von 04.2012 bis 03.2016.
09.2008 – 03.2011	Masterstudium an der Northwestern Polytechnical University, Xi'an, China Fachbereich: Leistungselektronik und elektrische Antriebe
09.2004 – 06.2008	Bachelorstudium an der Northwestern Polytechnical University, Xi'an, China Fachbereich: Elektrotechnik und Automatisierung

Berufserfahrungen

04.2011 – 03.2012	Elektroingenieurin, China Aerospace Science and Technology Corporation.

In der Reihe „*Forschungsberichte aus dem Lehrstuhl für Regelungssysteme*“, herausgegeben von Steven Liu, sind bisher erschienen:

1	Daniel Zirkel	Flachheitsbasierter Entwurf von Mehrgrößenregelungen am Beispiel eines Brennstoffzellensystems ISBN 978-3-8325-2549-1, 2010, 159 S. 35.00 €
2	Martin Pieschel	Frequenzselektive Aktivfilterung von Stromoberschwingungen mit einer erweiterten modellbasierten Prädiktivregelung ISBN 978-3-8325-2765-5, 2010, 160 S. 35.00 €
3	Philipp Münch	Konzeption und Entwurf integrierter Regelungen für Modulare Multilevel Umrichter ISBN 978-3-8325-2903-1, 2011, 183 S. 44.00 €
4	Jens Kroneis	Model-based trajectory tracking control of a planar parallel robot with redundancies ISBN 978-3-8325-2919-2, 2011, 279 S. 39.50 €
5	Daniel Görges	Optimal Control of Switched Systems with Application to Networked Embedded Control Systems ISBN 978-3-8325-3096-9, 2012, 201 S. 36.50 €
6	Christoph Prothmann	Ein Beitrag zur Schädigungsmodellierung von Komponenten im Nutzfahrzeug zur proaktiven Wartung ISBN 978-3-8325-3212-3, 2012, 118 S. 33.50 €
7	Guido Flohr	A contribution to model-based fault diagnosis of electro-pneumatic shift actuators in commercial vehicles ISBN 978-3-8325-3338-0, 2013, 139 S. 34.00 €

8	Jianfei Wang	Thermal Modeling and Management of Multi-Core Processors ISBN 978-3-8325-3699-2, 2014, 144 S. 35.50 €
9	Stefan Simon	Objektorientierte Methoden zum automatisierten Entwurf von modellbasierten Diagnosesystemen ISBN 978-3-8325-3940-5, 2015, 197 S. 36.50 €
10	Sven Reimann	Output-Based Control and Scheduling of Resource-Constrained Processes ISBN 978-3-8325-3980-1, 2015, 145 S. 34.50 €
11	Tim Nagel	Diagnoseorientierte Modellierung und Analyse örtlich verteilter Systeme am Beispiel des pneumatischen Leitungssystems in Nutzfahrzeugen ISBN 978-3-8325-4157-6, 2015, 306 S. 49.50 €
12	Sanad Al-Areqi	Investigation on Robust Codesign Methods for Networked Control Systems ISBN 978-3-8325-4170-5, 2015, 180 S. 36.00 €
13	Fabian Kennel	Beitrag zu iterativ lernenden modellprädiktiven Regelungen ISBN 978-3-8325-4462-1, 2017, 180 S. 37.50 €
14	Yun Wan	A Contribution to Modeling and Control of Modular Multilevel Cascaded Converter (MMCC) ISBN 978-3-8325-4690-8, 2018, 209 S. 37.00 €
15	Felix Berkel	Contributions to Event-triggered and Distributed Model Predictive Control ISBN 978-3-8325-4935-0, 2019, 185 S. 36.00 €
16	Markus Bell	Optimized Operation of Low Voltage Grids using Distributed Control ISBN 978-3-8325-4983-1, 2019, 211 S. 47.50 €

17	Hengyi Wang	Delta-connected Cascaded H-bridge Multilevel Converter as Shunt Active Power Filter	
		ISBN 978-3-8325-5015-8, 2019, 173 S.	38.00 €

Alle erschienenen Bücher können unter der angegebenen ISBN im Buchhandel oder direkt beim Logos Verlag Berlin (www.logos-verlag.de, Fax: 030 - 42 85 10 92) bestellt werden.